Marine Propellers and Propulsion

Marine Propellers and Propulsion

Editor

Alfred Eckstein

scitus academics

Marine Propellers and Propulsion
Edited by **Alfred Eckstein**

ISBN: 978-1-68117-574-4
Library of Congress Control Number: 2016943009

www.scitusacademics.com
Box No. 4766, 616 Corporate Way,
Suite 2, Valley Cottage,
NY 10989

Notice

Printed in United States of America on Acid Free Paper

Preface

Marine propellers and propulsion is in fact a complex, multidisciplinary topic with implications for design, construction, operation and research. A propeller is a rotating fan as a structure used to drive the ship using the power generated and transmitted by the ship's primary engine. The transmitted power is converted from rotational motion to generate a thrust that gives the water momentum, leading to a force acting on the ship and pushing it forward. Boats are able to maneuver them in the water with propulsion forces. Although there were initially limited numbers of ship propulsion systems, there are many innovative systems with which a boat can be equipped in the present era. The propulsion of the ship today is not just about good movement of the ship in water also includes the use of the best propulsion method to ensure a much better safety standard for the marine ecosystem and cost efficiency.

This book offers comprehensive and gradual coverage to marine engineers and anyone dealing with propulsion and hydrodynamics with the necessary knowledge to do the job. This book combines a wide range of propulsion technology knowledge, a multidisciplinary and international topic.

Table of Contents

1

CHAPTER

Hydro-Structure Analysis of Composite Marine Propeller under Pressure Hydrodynamic Loading

Hassan Ghassemi[1], Manouchehr Fadavie[1], Daniel Nematy[1]

[1]*Department of Ocean Engineering, Amirkabir University of Technology, Tehran, Iran*

ABSTRACT

This paper aims to predict the hydrodynamic characteristics and structural analysis of the marine propeller under pressure hydrodynamic loading. Because of the loading on the propeller blade, it goes under significant deformation that may affect the hydrodynamic performance of the propeller. Thus, the blade deformation of a propeller due to fluid pressure should be analyzed, considering hydro-elastic analysis. The propeller was made of anisotropic composite materials, and the geometry of the propeller is for one skew angle. First, the hydrodynamic pressure loading is obtained by FVM and then the deformation of the blade due to this pressure was calculated. Next, the pressure load for deformed propeller is achieved; it is again repeated to obtain the new deformed propeller. This procedure is repeated to converge the thrust, torque and efficiency. We present all results of the pressure distribution, hydrodynamic characteristics, stress and deformation of the propeller.

Keywords: hydrodynamic characteristics, structural deformation, pressure and stress, composite material

1. INTRODUCTION

Marine propellers are elements in which operate behind the vessel's hull and generate the desired thrust due to pressure loading on the blades. These propulsion systems are the most efficient driving equipment to drive the marine vessels. Meanwhile metal alloys were widely used from old days as conventional materials in manufacturing the propellers. Also, Nickel-Aluminum-Bronze and Manganese-Nickel-Aluminum-Bronze, because of their superior properties such as high strength, corrosion resisting and manufacturing, are alloys that mostly employed for propellers structure.

Nowadays composite materials applications are expanding in many industries such as aerospace, marines and many other vital vehicles. Composite materials have special properties such as light weight, strength-to-weight ratio, stiffens-to-weight ratio and easy manufacturing. Nowadays, extra enthusiasm for the use of composite materials in marine industries, in terms of improving properties and performance of structures, exists, especially in manufacturing the propeller blades as a result of anisotropic properties of these composite materials. Also, increase of efficiency and decrease of noise, fuel consumption, vibration, loading on blades, weight and pressure fluctuations are the common consequences of applying composite materials in marine propellers

According to increasing applications of composite materials, studying about these materials is needed more than ever. Due to complicated geometry of the propeller, structure analyzing is difficult. Many researchers have employed analytical and numerical method to determine the fluid-structure interactions of the propeller[1, 2]. Formerly, cantilever beam theory is used for majority of investigations. Furthermore because of complicated shape of the blade FEM is mostly used for structural analyzing of propellers.

For hydrodynamic analysis and calculation of the pressure distribution on the propeller blade several methods have been developed until now; such as blade element theory, lifting lines theory, vortex lattice method and boundary element method. Lin and Lin worked on nonlinear hydro-elastic behavior of propellers using a finite-element method and lifting surface theory [3]. Cho and Lee [4] published an article entitled propeller blade shape optimization for efficiency improvement. They studied an optimization method to generate the optimal shape of the propeller to increase the efficiency. Meanwhile lifting line theory and lifting surface theory are applied for hydrodynamic calculations.

A coupled BEM-FEM method for exploring the fluid structure interaction of flexible composite marine propellers is developed by Young [5, 6]. Also, Young and Motely investigated influence of material and loading effects on the propellers made of advanced materials [7]. Lin et al. published nonlinear hydroelastic behavior of propellers using a finite-element method and lifting surface theory [8] and also experimental paper which presented optimization and experiments on a composite propeller [9]. A FSI (fluid structure interaction) analysis for a flexible propeller is done by Sun et al. [10]. Kulczyk and Tabaczek studied the propeller-hull interaction in the propulsion system [11]. The application of composite materials on the ships propeller performance is investigated by Takestani et al. [12]. Also, other works of the hydro-elastic analysis of marine propellers have been performed in [13, 14].

The boundary element method (BEM) is widely used for analyzing the flow around the propellers. This method is applied in three-dimensional turbulence flow, based on potential theory. The BEM is developed for simulating the cavitation in face and back of the immersed, supercavitation and surface piercing propeller in wake flow by Kinnas and Fine [15]. Additionally, the accuracy of this method was so good for no vibration case, pressure and cavitation calculated by BEM was considerably close to experimental results. Ghassabzadeh and Ghassemi employed a hybrid BEM-FEM to determine the hydrodynamic characteristics of a composite propeller based on hydro-elastic analyzing [16]. Recently, investigation of different methods of noise reduction for submerged marine propeller s and their classification carried out by Feiz Chekab et al. [17].

In this paper, finite volume method is used to calculate the hydrodynamic pressure. On the other hand, the blade deformation is analyzed by FEM. Firstly, the hydrodynamic pressure for undeformed propeller is calculated by FVM and the results are exported to the FEM. Then the strain and stress of the blade is calculated. After that the pressure distribution on the propeller using the deformed shape of the blade is recalculated and the new pressure loading is exported to the FEM code and evaluate new deformation again. This process was repeated until the optimum shape and results are obtained.

2. FVM-FEM COUPLED

The finite volume method with coupled finite element method employed to the present calculations. Pressure on the blade of the propeller and hydrodynamic characteristic of the propeller are first calculated by FVM. Deformed blades and stress results may be determined by finite element method.

Here, the computational results of thrust and torque compared to experimental results [16] for B-series propeller are shown in Figure 1, which examine a very good agreement between them. In other word, the numerical results based on experimental data were proved.

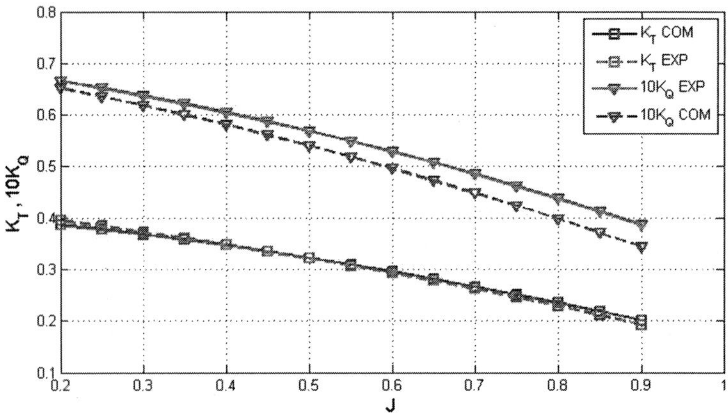

Figure 1. Comparison of computational and experimental hydrodynamic characteristic

Geometry and Computation Grid

In this paper, a marine propeller for B-series was chosen and modeled. The results of hydrodynamic and structural analyzing were presented. The process is according to the flowchart shown in Figure 2. The propeller geometric properties were also shown in Table 1, given from reference [16].

Table 1. Propeller Geometry Properties

parameter	Value
Number of Blades	5
Diameter	0.5 m
Hub Ratio	0.2
Pitch Ratio	Variable (at 0.7R=0.8)
Expanded Area Ratio (EAR)	0.65
Skew Angle	Variable (at 0.7R=25 degree)
Rake Angle	10 degree

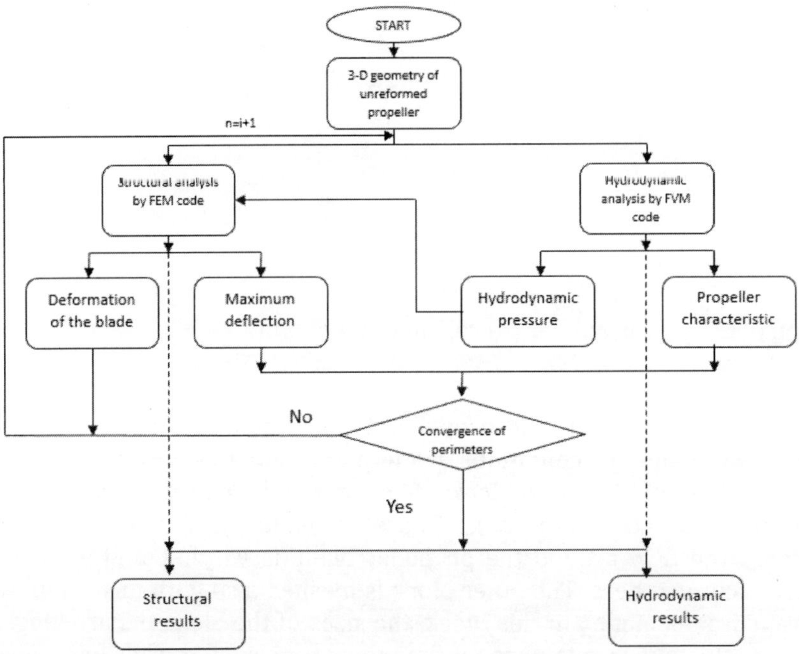

Figure 2. Calculations flowchart

At the beginning of the process, the propeller geometry and domain is generated as shown in Figure 3. The Cartesian coordinates system is used, where x,y and z denote downstream, starboard and upward direction respectively. The origin was located at the center of the hub, and positive directions are upstream, starboard and downstream. The domain dimensions were shown in Figure 4. The solution field was divided into global and sub-domain. The sub-domain frame simulates the propeller rotation and employs the Coriolis acceleration terms in the governing equations for the fluid. The global frame surrounds the sub-domain frame. The global frame is a circular cylinder with 4D diameter, where D is the propeller diameter. The distance between the sub-domain frame and inlet is nearly 4D, while it is nearly 6D for the outlet and dynamic frame. The sub-domain is sphere with a diameter of D.

Figure 3. (a) 3D propeller (b) calculation domain

Discretized domain around propeller is illustrated in Figure 5. In the next step, boundary conditions are imposed. For this case we divided the solution field into two blocks. The rotating block was meshed with unstructured tetrahedral cells. These elements have a smaller size because the flow around the propeller blade is varying, and we need more accuracy there. The other block is meshed also with unstructured tetrahedral elements. In this block, the sizes of the elements are bigger because the flow in this region is smoother, and smaller grids just cause longer calculation procedure. We employed different grid number and it is concluded that with number of elements 991640 are sufficient for the present calculations.

Table 2. Mechanical Properties of Composite and Coper High Tensile Brase

Property Material	Modulus of elasticity (GPa)	Poisson ratio	Shear modulus elasticity (GPa)	Density (ton/m3)
Composite	E_x=132 E_y=10.8 E_z=10.8	v_{xy}=0.24 v_{yz}=0.49 v_{xz}=0.24	G_{xy}=5.65 G_{yz}=3.38 G_{xz}=5.65	ρ_F=1.5
Copper High Tensile Brass	E_x=102.97	v_{xz}=0.35	-	ρ_F=8.25

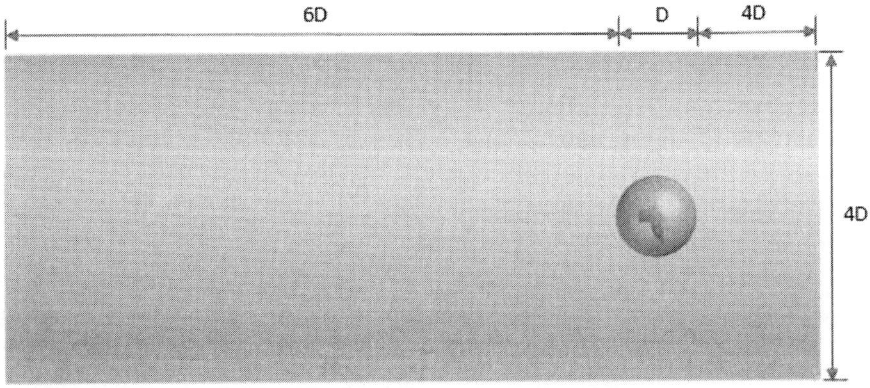

Figure 4. Domain dimensions

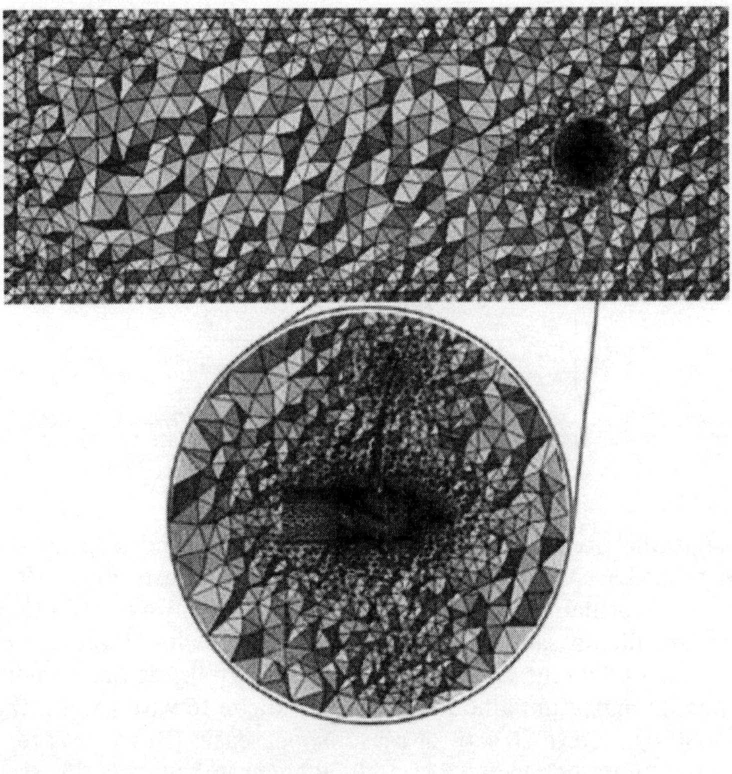

Figure 5. Domain descritization

Boundary Conditions and Numerical Setups

In the present paper, we used CFX solver in order to evaluate the hydrodynamic performance of a B-series propeller. The governing equations of this problem were solved by the finite volume method based on the RANS equations. A SIMPLE algorithm was used for solving the pressure-velocity coupling equations. The k-epsilon model was used for turbulence model. The fluid (water) was assumed as an incompressible fluid, therefore; on the inlet boundary, uniform velocity is imposed and on the outlet boundary, pressure outlet was imposed. The axial velocity is 0.2m/s in the opposite direction. The no-slip wall condition was imposed on the propeller blades.

ANSYS structural 14 used for structural analyzing. The solver type was iterative. Propeller blade was made of linear elastic anisotropic composite with mechanical properties as shown in Table 2. We also define a fixed support at the end of the hub.

Hydrodynamic and Structural Results

The performance characteristics of the marine propeller can be defined using the non-dimensional coefficients such as the advance ratio, thrust coefficient, torque coefficient, and efficiency, which can be computed, respectively, as follows:

$$\begin{cases} J = \dfrac{V}{nD}, & K_T = \dfrac{T}{\rho n^2 D^4} \\ K_Q = \dfrac{Q}{\rho n^2 D^5}, & \eta = \dfrac{J}{2\pi}\dfrac{K_T}{K_Q} \end{cases} \quad (1)$$

After solving, the pressure contour, at J=0.2, in face and back of the deformed propeller is shown in Figure 6 and Figure 7, respectively. It is clear that the deformation of the blade affects the pressure distribution. The hydrodynamic pressure of the deformed blade is less than that of the undeformed blade. As can be seen the pressure in regions near to the root has its maximum amount and as we move toward the tip the pressure loading, due to structural reasons, decreases. However, areas near to mid-sections, between 0.5R and 0.7R generate the most thrust.

Figure 6. Pressure contour in face

Figure 7. Pressure contour is back

Figure 8. Deformation field of the propeller blade

Figure 8 shows the deflection of the blade along the radial at J=0.2. From the hydrodynamic point of view, applied loading on the blade is the heavy condition that may act to reach high deformation. As can be seen, deflections are distributed according to the nonlinear function. Blade is deformed higher at the tip. Maximum deflection is about 6 cm at the tip. The reason of this phenomenon is that a major thrust was generated in middle sections of the blade and also as we move toward the tip the bending moment about the root increases. Also, if deformation of the blade monitored carefully, the deformation of the trailing edge is more than of the leading edge. There is also an explanation for that, the stress, which illustrated in following Figures, is higher in the trailing edge.

The deformed and undeformed shape of the blade were shown with ten times enlargement at J=0.2, in Figure 9. As mentioned earlier, the deformation increases from root to tip of the blade, due to structural reasons mentioned earlier. The vectors of deformation or one blade is shown in Figure 10.

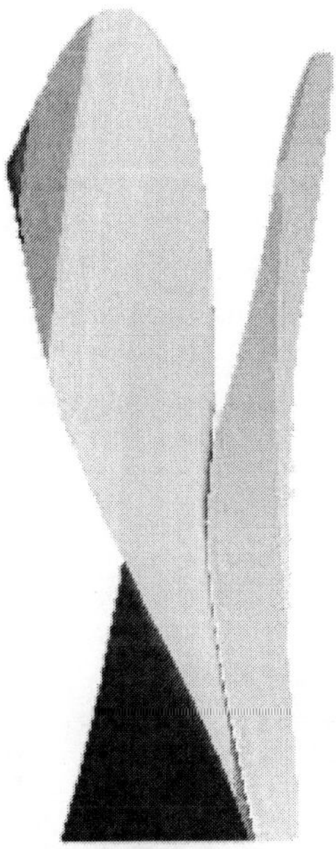

Figure 9. Deformed and undeformed blade of the propeller

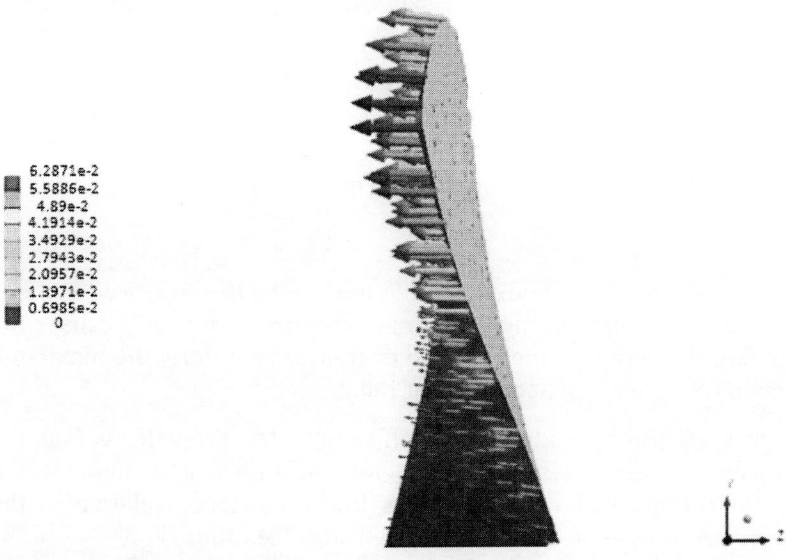

Figure 10. Deformation vectors of the blade

The equivalent von-misses stress in the face and back of the propeller was shown in Figs. 11 and 12 respectively. The stress in the trailing edge has its maximum value and as we move towards the other sections of the blade the equivalent stress decrease. The stress in tip is about 22% of maximum equivalent stress and in the leading edge in about 66.5% of maximum equivalent stress. In the back of the propeller in middle sections, we have the maximum equivalent stress and as we move to the other blade sections the stress decrease.

Figure 11. Equivalent stress in the face of the propeller

Figure 13 shows the maximum deflection versus the advanced velocity ratio; it is clear that, an increase in advanced velocity ratio causes the maximum deflection of the blade to decrease. Therefore, the maximum deflection for the propeller occurs at J=0.2.

In Figure 14, the hydrodynamic coefficient of the propeller versus the advanced coefficient is presented for deformed and undeformed propellers. Deformed propeller gives lower efficiency relative to the undeformed propeller, especially at high advance ration.

Figure 12. Equivalent stress in the back of the propeller

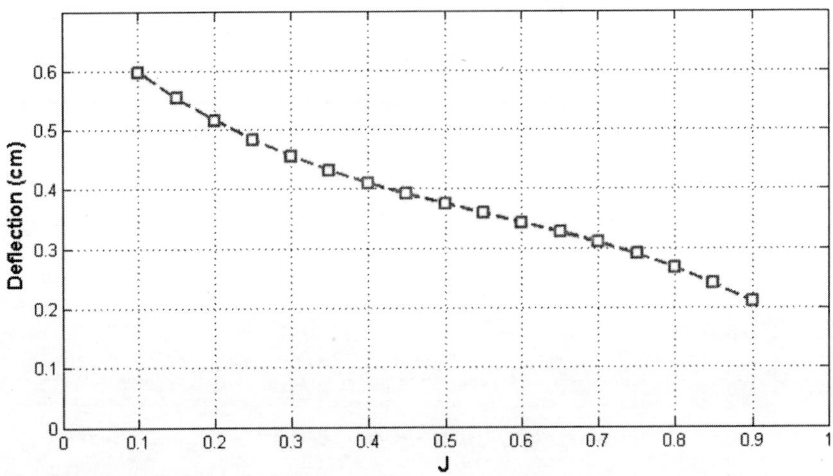

Figure 13. Deflection of the blade versus advanced coefficient

In order to examine the effect of the advanced velocity ratio on the hydrodynamic characteristics precisely, the relative error of the propeller characteristics relative to those of the undeformed blade are shown in Figure 15 at all advanced velocity ratio. It is clear that relative error in the propeller characteristics are nearly constant, and so the performance of the propeller is independent of the rotational speed and the advance velocity.

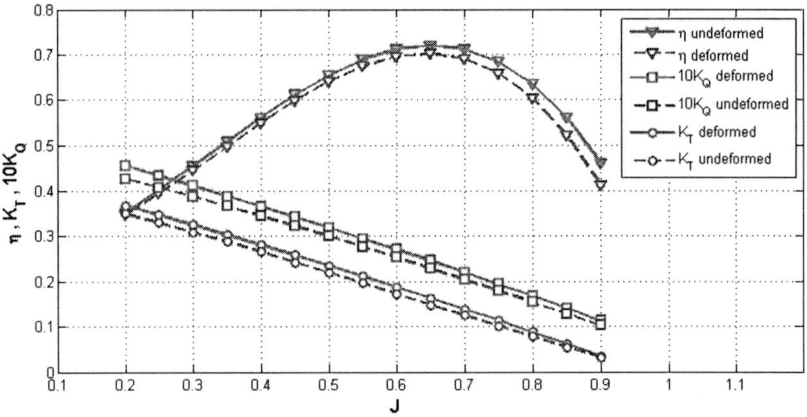

Figure 14. Hydrodynamic coefficients of deformed and undeformed propeller

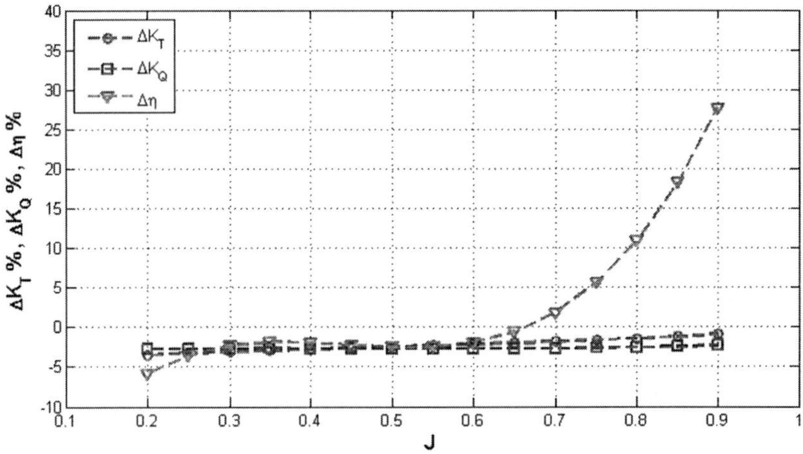

Figure 15. Relative error in hydrodynamic coefficients

3. CONCLUSIONS

In the present paper, coupled FVM-FEM hydro-elastic procedure was used to investigate the hydrodynamic and structure behavior of a marine propeller. The solving procedure was an iterative process; the hydrodynamic pressure was calculated in FVM code and the structural analysis performed by FEM code. In the case study, a five blade composite propeller was chosen to study the deformations and the effect of deformation on the propeller performance. Admittedly, the deformation toward the blade tip increases and stress decreases. The stress in the face of the blade and leading edge is maximum and in back and in the middle of the blade, and as moves to the other areas decrease. Increasing in advanced velocity ratio causes the maximum deflection, the thrust coefficient and the torque coefficient decreases.

We also determined the effect of this deformation on the propeller hydrodynamic performance. Obviously, in low advanced coefficients, when the propeller is working in heavy conditions, both thrust force coefficient and torque coefficient increase. While the propeller performance decrease. Hydrodynamic coefficients of the propeller are approximately constant. That means that the deformation of the propeller is independent of the advance velocity ratio.

ACKNOWLEDGMENT

This research was supported by the Marine Research Center (MRC) of Amirkabir University of Technology (AUT) whose works are greatly acknowledged.

REFERENCES

1. Blasques, J. P., Berggreen, C., and Andersen, P. (2010). Hydro-elastic analysis and optimization of a composite marine propelle. Marine Structures, 23 (1), 22-38.

2. Cho, J., and Lee, S.-C. (1998). Propeller Blade Shape Optimization for Efficiency Improvment. Computers andFluids, 27 (3), 407-419.

3. Li, G., Li, W., You, Y., and Yang, C. (2013). Study on Fluid-Structure Interaction Characteristics of Composite Marine Propeller. In The Twenty-third International Offshore and Polar Engineering Conference. Anchorge: International Society of Offshore and Polar Engineers.

4. Lee, H., Song, M. C., Suh, J. C., and Chang, B. J. (2014). Hydro-elastic analysis of marine propellers based on a BEM-FEM coupled FSI algorithm. International Journal of Naval Architecture and Ocean Engineering, 6 (3), 562-577.

5. Young, Y. L. (2007). Fluid–structure Interaction Analysis of Flexible Composite Marine Propellers. Journal of fluids and structures.

6. Young, Y. L. (2006). Numerical and experimental investigations of composite marine propellers. Proceedings of Twenty-Sixth Symposium on Naval Hydrodynamics. Rome.

7. Young, Y. L., and Motley, M. R. (2011). Influence of Material and Loading Uncertainties on the Hydroelastic Performance of Advanced Material Propellers. Second International Symposium on Marine Propulsors, (pp. 10-17). Hamburg.

8. Lin, H. J., and Lin, J. J. (1996). Nonlinear hydroelastic behavior of propellers using a finite-element method and lifting surface theory. Journal of Marine Science and Technology, 1 (2), 114-124.

9. Lin, C. C., Lee, Y. J., and Hung, C. S. (2009). Optimization and experiment of composite marine propellers.Composite Structures, 89 (2), 206-215.

10. Sun, H. T., and Xiong, Y. (2012). Fluid-structure interaction analysis of flexible marine propellers. Applied Mechanics and Materials, 226 (228), 479-482.

11. Kulczyk J. and Tabaczek T. (2014) Coefficients of Propeller-hull Interaction in Propulsion System of Inland aterway Vessels with Stern Tunnels, the International Journal on Marine Navigation and Safety of Sea Transportation, Vol. 8 No. 3.

12. Takestani, T., Kimura, K., Ando, S., and Yamamoto, K. (2013). Study on Performance of a Ship Propeller Using a Composite Material. 3rd International Symposium on Marine Propulsors. Launceston, Tasmania.

13. He, X. D., Hong, Y., and Wang, R. G. (2012). Hydroelastic optimisation of a composite marine propeller in a non-uniform wake. Ocean Engineering, 39, 14-21.

14. Mulcahy, N. L., Prusty, B. G., and Gardiner, C. P. (2010). Hydroelastic tailoring of flexible composite propellers.Ships and Offshore Structures, 5 (4), 359-370.

15. Kinnas, S. A., and Fine, N. E. (1993). A Boundary Element Method for the Analysis of the Flow Around 3-d Cavitating Hydrofoils. Journal of Ship Research, 3 (37), 213-224.

16. Ghassabzadeh, M., Ghassemi, H., and Saryazdi, M. G. (2013). Detrmination of Hydrodynamics Characteristics of Marine Propeller Using Hydroelastic Analysis. Brodogradnja, 64 (1), 40-45.

17. Feizi Chekab M., Ghadimi P., Djeddi, S.R., Soroushan M., Investigation of Different Methods of Noise Reduction for Submerged Marine propeller s and Their Classification, American Journal of Mechanical Engineering. 2013 1 (2).

2

CHAPTER

Hydro-Elastic Analysis of Marine Propellers Based on a BEM-FEM Coupled FSI Algorithm

Hyoungsuk Lee[1] , Min-Churl Song[2] , Jung-Chun Suh[3] and Bong-Jun Chang[1]

[1] *Maritime Research Institute, Hyundai Heavy Industries, Co., Ltd., Ulsan, Korea*

[2] *Industrial Technology Institute, Hyundai Heavy Industries, Co., Ltd., Ulsan, Korea*

[3] *Dept. of Naval Architecture and Ocean Engineering & Research Institute of Marine Systems Engineering, Seoul National University, Seoul, Korea*

ABSTRACT

A reliable steady/transient hydro-elastic analysis is developed for flexible (composite) marine propeller blade design which deforms according to its environmental load (ship speed, revolution speed, wake distribution, etc.) Hydro-elastic analysis based on CFD and FEM has been widely used in the engineering field because of its accurate results however it takes large computation time to apply early propeller design stage. Therefore the analysis based on a boundary element method-Finite Element Method (BEM-FEM) Fluid-Structure Interaction (FSI) is introduced for computational efficiency and accuracy. The steady FSI analysis, and its application to reverse engineering, is designed for use regarding optimum geometry and ply stack design. A time domain two-way coupled transient FSI analysis is developed by considering the hydrodynamic damping ffects of added mass due to fluid around the propeller blade. The analysis makes possible to evaluate blade strength and also enable to do risk assessment by estimating the change in performance and the deformation depending on blade position in the ship's wake. To validate this hydro-elastic analysis methodology, published model test results of P5479 and P5475 are applied to verify the steady and the transient FSI analysis, respectively. As the results, the proposed steady and unsteady analysis methodology gives sufficient accuracy to apply flexible marine propeller design.

Keywords: Marine propeller; Composite material; Fluid-structure interaction (FSI); Boundary element method (BEM, Panel method); Finite element method (FEM);

INTRODUCTION

Research background

A marine propeller operates in the complex flow field called the ship's wake and the transition of flow angle of attack leads to rapid changes in pressure distribution on the propeller blade. As the trend of high power engines follows increases in vessel size, a marine propeller is being exposed to various risks including cavitation erosion, pressure fluctuation-induced vibration, underwater radiation noise, etc.

Compared with propellers made of metal, those made of composite materials - a combination of two or more physically/chemically heterogeneous materials - have superior properties for managing these risks. Composite materials are light weight, and have high stiffness and better sound absorption ability (ITTC, 2005). Despite their higher costs, the excellent manufacturing and maintenance ability of an additive production process is expanding the application of composite materials to all industrial areas, including aviation, environment/energy, information, electricity, automobile and household supplies.

Attempts are being made to use composite materials in the marine propeller blades of warships, submarines, and torpedoes for two reasons. First, composites can improve chance of survival because their good sound absorption ability reduces noncevitating propeller noise. Second, their blade deformation characteristics of the wake distribution delays cavitation inception and reduces cavitation volume. Their superior properties have led to their expanded application in yachts, tugs, and fishing boats. Composite materials also make it possible to reduce A/S costs for preventing heat damages in shaft systems, and to simplify the vessel stern structures due to their light weight - less than half the weight of Nickel-Aluminum-Bronze (NAB) alloy (ITTC, 2005).

The amount and geometry of a composite material's deformation from the acting load can be controlled with different combinations of laminated material, stacking order, and ply angle. Thus, if its deformable characteristics can be properly considered at the design stage, a composite propeller can reduce fuel oil consumption in its operating profile compared with an NAB propeller which does not change shape (Young, 2007a; 2007b; 2008).

While recognizing these virtues, the difficulty is that it is not easy to accurately and practically simulate the hydro-elastic behavior of the composite marine propellers. These simulation difficulties act as a barrier to the accurate prediction of propulsion performance, initial geometry decisions, and strength evaluations at the propeller design stage.

Previous studies

Inviscid numerical methodologies for calculating the flow field around propeller blades include the lifting line method, the lifting surface method (or vortex-lattice method, VLM), and the boundary element method (BEM). These inviscid methods are widely used in propeller performance prediction because of their short computing times and general usability. Computational fluid dynamics (CFD) analyses (RANS, LES, etc.) that consider the viscous effect tend to use a propeller flow analysis, which requires increased computing power. Even though the viscous flow analyses give a more accurate solution, BEM-based FSI analysis is performed in this study due to the considerations of computing time for the evaluation of propeller design applications.

The lifting line theory, proposed by Prandtl (1921), and the VLM, which was applied to the propeller by Kerwin et al. (1978), are useful methods for initial propeller design due to their ease of calculation. They have limitations, however, when considering three-dimensional blade geometry, BEM is introduced to achieve a more accurate FSI analysis. BEM, also called the panel method, was adapted to the propeller by dividing the potential basis and velocity basis methods of Hess and Valarezo (1985), Lee (1987),Kerwin et al. (1987), Hoshino (1989), etc. After Lee (1987) showed the excellent performance of the potential basis method, this method has been mainly used in the field of propeller analysis. Hsin (1990) and Suh et al. (1992) proposed the Kutta condition and Greeley and Kerwin (1982) and Lee (1987) introduced the flow-adapted grid to include a propeller wake roll-up model for improved convergence in the BEM analysis. These methods apply to the BEM code, KPA14, used in this study.

Various other numerical and experimental studies have also been conducted to analyze composite material (or elastomer) propellers. Lin and Lin (1996) simulated the elastic behavior of marine propellers in noncavitating conditions with a VLM-FEM interaction.Young (2007a; 2008) showed a hydro-elastic response of a composite propeller in steady/unsteady conditions by using BEM-FEM FSI and validated the results with model test results. In particular, Young's study conducted BEM analysis to predict the hydrodynamic damping effect for unsteady behavior such as blade "fluttering" induced by a change in inflow conditions at the hull behind the wake field. Chen et al. (2006)evaluated the behavior of the composite marine propeller in four-cycle inflow conditions by performing model tests and a numerical simulation. Motley et al. (2009) and Blasques et al. (2010) studied optimum material stacking to improve propulsion efficiency and Jang et al. (2012) proposed a reverse engineering procedure for the initial design of an elastomer propeller considering blade deformation. Lee et al. (2009), Lee et al. (2012)and Lee et al. (2013) have carried out the study of the relationship between manufacturing and performance of flexible (composite) propeller.

Objective

The hydro-elastic analysis of a flexible marine propeller, which is a simplified orthotropic-homogeneous material model of a composite propeller, has been developed to accurately and efficiently predict the propeller's propulsion performance and evaluate its strength. For practical use in the propeller design stage, the inviscid potential-based BEM code, KPA14, is used to calculate the loads on the blades. A commercial Finite Element Analysis (FEA) solver (Abaqus Standard) is introduced for predicting blade deformation and evaluating propeller strength. An interface between these two solvers is developed to analyze the hydro-elastic behavior of the composite-material propeller, and the accuracy and usability of the developed algorithm are confirmed by comparing our results with the model test results in previously published literature.

FORMULATION AND NUMERICAL METHODS

The flow field around the propeller blade, the structural deformation, and their relationship at the blade surface are numerically formulated for the FSI analysis to predict the performance of the marine propeller.

Governing equation of the flow field around the propeller blades

As mentioned in the previous section, the potential basis analysis, from the studies ofHsin (1990), Suh et al. (1992), Greeley and Kerwin (1982) and Lee (1987), are applied to analyze the steady or unsteady flow fields. We assume that the flow field consists of the three-dimensional closed boundary S and the unit normal vector $\mathbf{n}$ toward the flow to formulate the flow around the propeller blades. The boundary S is composed of the blade and hub S_B, the wake S_W, and the outer boundary S_∞. The propeller is rotating in the incompressible, inviscid, and irrotational flow with inflow velocity $\mathbf{U}_\infty$. The general notation of object in fluid for the application of Green's theorem is shown in Fig. 1.

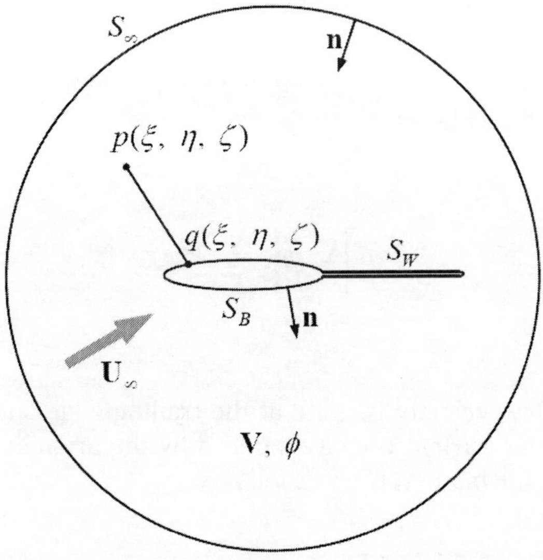

Figure 1. General notation of object in fluid for the application of green's theorem.

The perturbed velocity potential ϕ satisfies the Laplace equation.

$$\nabla^2\phi = 0 \tag{1}$$

When applying the boundary condition to the boundaries, the boundary value problem can be formulated and the kinematic boundary condition must be satisfied on the propeller blade surface S_∞ as follows:

$$\frac{\partial\phi}{\partial n} = -\mathbf{U}_\infty \cdot \mathbf{n} \tag{2}$$

Assume that wake surface S_W has no thickness and there is no normal velocity jump, then

$$\mathbf{n}\cdot\left(\mathbf{V}^+ - \mathbf{V}^-\right) = 0 \quad \text{on } S_W \tag{3}$$

The Kutta condition of Eq. (4) is applied to the unique circulation at the trailing edge of the blades.

$$\left|\nabla\phi\right|_{T.E.} < \infty \tag{4}$$

Namely, the flow velocity is finite at the trailing edge and at the outer boundary, the perturbed velocity induced by the propeller is vanishing far away from the blade surface as follows:

$$\nabla\phi \rightarrow 0\text{, as } \quad r \rightarrow \infty \tag{5}$$

The disturbed velocity potential ϕ can be expressed as Eq. (6) according to the governing equation because it should satisfy Green's 2nd identity for all points on the blade surface.

$$2\pi\phi_p = \iint_{S_B}\left[\phi_q \frac{\partial}{\partial n_q}\left(\frac{1}{R(p;q)}\right) - \left(\frac{\partial\phi}{\partial n}\right)_q \frac{1}{R(p;q)}\right]ds + \iint_{S_W}(\Delta\phi)_q \frac{\partial}{\partial n_q}\left(\frac{1}{R(p;q)}\right)ds \tag{6}$$

where

$S_\infty S_{blade} + S_{hub} p$ point induced velocity potential is calculated q

point positioned singularity

$R(p;q)$

distance between p and q

The geometry of the propeller blades, hub, and wake could be discretized as a panel to numerically solve the governing equation in Eq. (7).

$$\sum_j a_{i,j}\phi_j + \sum_m (\Delta\phi)_m W_{i,m} = \sum_j b_{i,j}\sigma_j \tag{7}$$

The effective coefficients $a_{i,j}$ and $b_{i,j}$ are the velocity potentials at the ith panel induced by the unit strength dipole and the source distributed at the jth panel, respectively. The wake effective coefficient $W_{i,m}$ is the velocity potential at the ith panel induced by the unit strength dipole at the mth strip at the wake panel. $(\Delta\phi)m$ is the difference in potential between the upper and lower side of the trailing edge and σ_j is the source strength on the blades. The diagram of discritized surface panel of blade, hub and wake and coordinate system is shown in Fig. 2.

To solve the unsteady transient flow problem defined by the time domain, a steady analysis for the key blade is performed for every time step and the singularity of the other blades is updated for every step shown in Eq. (8)

$$\sum_j a_{i,j}\phi_j(n) + \sum_m W_{i,m,1}(\Delta\phi)_{m,1}(n) = B_i(n) \tag{8}$$

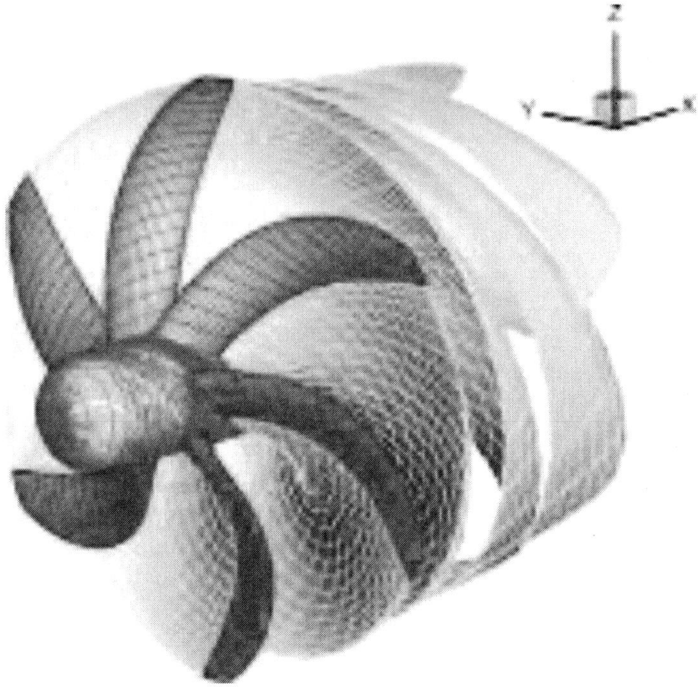

Figure 2. Discretized surface panel of blade, hub and wake and coordinate system.

$$B_i(n)=\sum_{k=1}\sum_{j} b_{i,j}^k \sigma_{i,j}^k(n)-\sum_{k=2}\sum_{j} a_{i,j}^k \phi_{i,j}^k(n)-\sum_{k=2}\sum_{m}\sum_{l=1} W_{i,m,l}^k\left(\Delta\phi\right)_{m,l}^k(n)-\sum_{m}\sum_{l=2} W_{i,m,l}\left(\Delta\phi\right)_{m,l}(n)$$

and variable *n* means time step, superscript *k* means the indices of blades, i.e. *k*=1 means the key blade.

The pressure acting on the blade surface can be obtained by using the velocity potential derived from the discretized integral equation in Eqs. (7) or (8) as follows:

$$C_p=\frac{p-p_\infty}{\frac{1}{2}\rho U_\infty^2}=-\frac{2}{U_\infty^2}\frac{\partial'\phi}{\partial t}+\frac{V_F^2}{U_\infty^2}-\frac{\left|\nabla\phi\right|^2}{U_\infty^2}-\frac{2V_F\cdot\nabla\phi}{U_\infty^2} \tag{9}$$

where the operator $\partial/\partial t$ in the inertial frame into $(\partial'/\partial t - V_F\nabla)$ in the moving frame.

Structural equations of motion for blade deformation

The equation of motion for the structural deformation corresponding to the propeller blade fixed coordinate is as Eq. (10),

$$\mathbf{M}_S\ddot{\mathbf{d}} + \mathbf{C}_S\dot{\mathbf{d}} + \mathbf{K}_S\mathbf{d} = \mathbf{F}_{\mathrm{ST}} \tag{10}$$

where $\mathbf{M}_s$ is the mass matrix, $\mathbf{C}_s$ is the damping matrix, $\mathbf{K}_s$ is the stiffness matrix which represents the material mass, damping and stiffness of the composite propeller blade. The values of mass, damping and stiffness in the validation have use of published experimental results. $\ddot{\mathbf{d}}$, $\dot{\mathbf{d}}$, $\mathbf{d}$ are acceleration, velocity, and displacement for the computing point, respectively.

$\mathbf{F}_{\mathrm{ST}}$ in Eq. (11) represents the total of all loads acting on the blade structure. These consist of the hydrodynamic pressure load $\mathbf{F}_{\mathrm{hp}}$ and the body force due to rigid body rotation, the Coriolis force $\mathbf{F}_{\mathrm{cori}}$ and the centrifugal force $\mathbf{F}_{\mathrm{cent}}$, and the coupling component force $\mathbf{F}_{\mathrm{fs}}$ acting on the interface between the fluid and the structure. The summation of the structural loads except $\mathbf{F}_{\mathrm{fs}}$ can be written as $\mathbf{F}_{\mathrm{s}}$.

$$\mathbf{F}_{\mathbf{ST}} = \mathbf{F}_{\mathbf{hp}} + \mathbf{F}_{\mathbf{cori}} + \mathbf{F}_{\mathbf{cent}} + \mathbf{F}_{\mathbf{fs}} = \mathbf{F}_{\mathbf{s}} + \mathbf{F}_{\mathbf{fs}} \tag{11}$$

The structural analysis of the hydro-elastic response is conducted with the commercial FEA solver Abaqus 6.12. Since the implicit method is adapted for both the steady and unsteady analyses, the static option of the Abaqus Standard is used for the steady state analysis and the dynamic option is used for the transient analysis.

The Newmark integration method of Eqs. (12) and (13) is applied to solve the nonlinear dynamic response problem and the model response at time $t + \Delta t$.

$$u_{t+\Delta t} = u_t + \Delta t\dot{u}_t + \Delta t^2\left[\left(\frac{1}{2} - \beta\right)\ddot{u}_t + \beta\ddot{u}_{t+\Delta t}\right] \tag{12}$$

$$\dot{u}_{t+\nabla t} = \dot{u}_t + \Delta t[(1-\gamma)\ddot{u}_t + \gamma\ddot{u}_{t+\Delta t}] \tag{13}$$

where parameters β and γ are the value-related stability and the accuracy of the convergence of analysis, respectively. 0.25 and 0.5 are applied for each parameter, providing second-order analysis accuracy.

The 8-node continuum shell element is used in the structural analysis. This element can reflect the three-dimensional geometry of the propeller blade and has the merit of giving a shell element to the plate-like structure, as well as having automatic mesh generation and ease in applying the material properties according to the ply.

Fluid-structure interaction

Assuming that deformation of the blade for the inflow variation is relatively large and that the change in the hydrodynamic characteristics due to geometrical alteration is not small, the FSI effect must be applied for the estimation on each hydrodynamic force and structural deformation. We propose using a two-way coupled FSI analysis that includes structural nonlinear analysis because the amount of blade deformation is not small to assume linear according to Young (2008). To analyze strictly for the elastically vibrating response due to the changing flow field in the ship's wake, the "wet" condition (considering not only hydrodynamic pressure distribution on blade but including hydrodynamic added mass/damping effect) must be considered. The acoustic fluid medium finite element model is adapted to consider this effect. This acoustic medium model was applied to ship hull vibration in contact with water by Paik (2010).

Let the fluid field around propeller blade be compressible and inviscid then the equation of equilibrium is as follows:

$$\nabla_p + \gamma\dot{\mathbf{v}} + \rho\ddot{\mathbf{v}} = 0 \tag{14}$$

where p is the hydrodynamic pressure, ρ is the density of fluid, γ is the volumetric constant (volumetric drag) for dissipation, and v and $\ddot{\mathbf{v}}$ represent velocity and acceleration of fluid particles, respectively.

Fluid volumetric strain can be represented as in Eq. (15) by the definition of the relationship between the bulk modulus and pressure.

$$p = -K_f \nabla \cdot \mathbf{v} \tag{15}$$

The equation of motion for hydro pressure can be derived from Eqs. (14) and (15).

$$\frac{1}{K_f}\ddot{p} + \frac{\gamma}{\rho K_f}\dot{p} - \frac{1}{\rho}\nabla^2 p = 0 \tag{16}$$

The equation of motion that considers the coupling of the boundary between the fluid and the structure can be expressed as the vector form.

$$\mathbf{M}_{\mathbf{f}}\ddot{\mathbf{p}} + \mathbf{C}_{\mathbf{f}}\dot{\mathbf{p}} + \mathbf{K}_{\mathbf{f}}\mathbf{p} - \mathbf{S}_{\mathbf{fs}}\ddot{\mathbf{d}} - 0 \tag{17}$$

where $\mathbf{M}_f$, $\mathbf{C}_f$, $\mathbf{K}_f$ and $\mathbf{S}_{fs}$ are $1/K_f$, $\gamma/(\rho\ K_f)$, $-1/\rho\ \Delta^2$ and coupling matrix respectively. And **p**represents a pressure vector in the fluid domain, **d** represents a unit vector on the fluid-structure boundary. Eq. (18) is derived from Eqs. (10) and (11) for structural deformation and then the system equation of FSI (Eq. 19) is obtained by substitution with Eq. (17).

$$\mathbf{M}_s\ddot{\mathbf{d}} + \mathbf{C}_s\dot{\mathbf{d}} + \mathbf{K}_s\mathbf{d} + \mathbf{S}_{fs}^T\mathbf{p} = \mathbf{F}_s \tag{18}$$

$$\begin{bmatrix} \mathbf{M}_s & 0 \\ -\mathbf{S}_{fs} & \mathbf{M}_f \end{bmatrix}\begin{Bmatrix} \ddot{\mathbf{d}} \\ \ddot{\mathbf{p}} \end{Bmatrix} + \begin{bmatrix} \mathbf{C}_s & 0 \\ 0 & \mathbf{C}_f \end{bmatrix}\begin{Bmatrix} \dot{\mathbf{d}} \\ \dot{\mathbf{p}} \end{Bmatrix} + \begin{bmatrix} \mathbf{K}_s & \mathbf{S}_{fs}^T \\ 0 & \mathbf{K}_f \end{bmatrix}\begin{Bmatrix} \mathbf{d} \\ \mathbf{p} \end{Bmatrix} = \begin{Bmatrix} \mathbf{F}_s \\ 0 \end{Bmatrix} \tag{19}$$

The tetra mesh (AC3D4 known as an element type of Abaqus Standard) of the acoustic medium domain is composed to perform structural analysis with consideration to the hydrodynamic damping due to added mass are shown in Fig. 3.

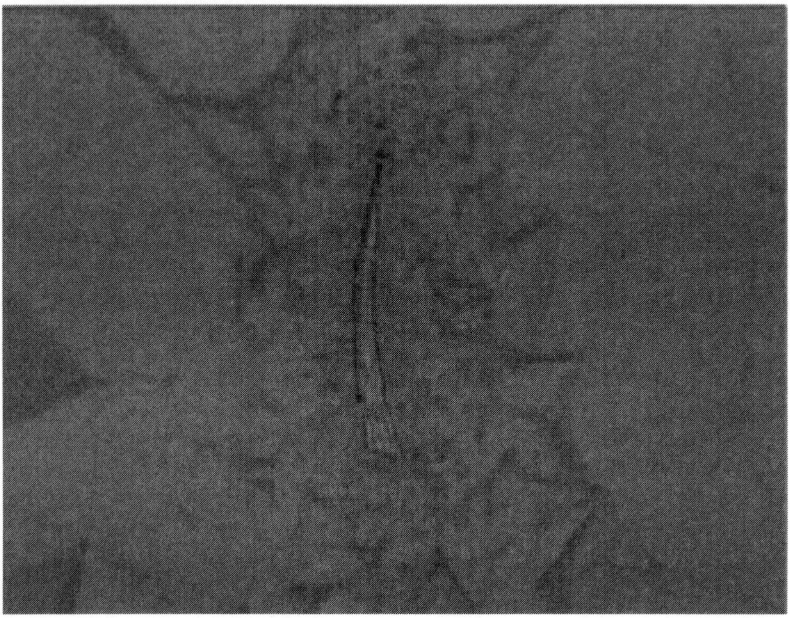

Figure 3. Grid system for structural analysis in acoustic medium ('wet' condition).

METHODOLOGY FOR HYDRO-ELASTICTY ANALYSIS OF A PROPELLER BLADE

Based on the formulated mathematical model, we implement a steady and unsteady transient FSI algorithm to predict the performance and evaluate the structure of the flexible propeller.

Basic concept of FSI analysis

A basic interface structure of the two-way coupled FSI, shown in Fig. 4, consists of a BEM panel code (KPA14) for calculating the hydrodynamic force acting on the blade surface and a nonlinear FEM solver (Abaqus Standard) for calculating the structural deforming response, rotational body force (Coriolis force and centrifugal force), and added mass damping of the unsteady transient flow condition.

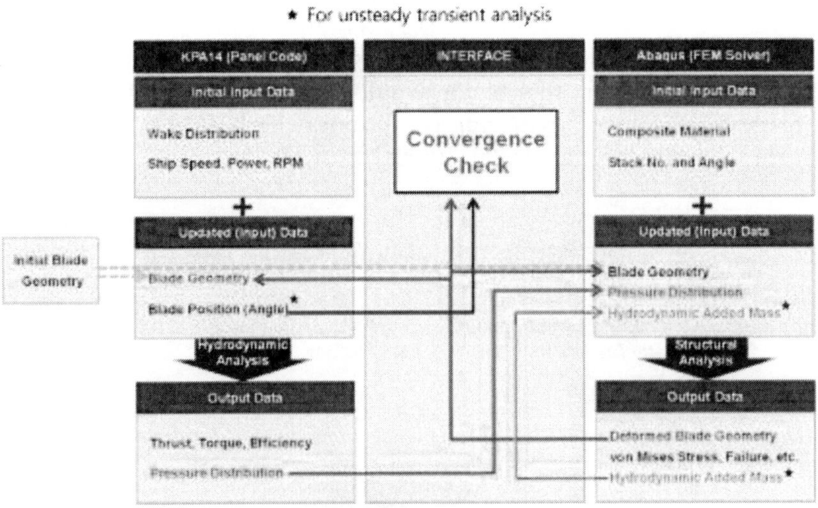

Figure 4. Schematic diagram for propeller blade FSI.

In steady-state FSI analysis, it is not necessary to consider the change in pressure distribution and the blade fluttering effect according to the blade position, so * items are excluded. The convergence is determined by the progress of comparison of geometrical differences from the former step (steady) and the former revolution (transient). Thus it is considered as converge when the maximum difference of nodal point is less than 0.01% of propeller diameter in the iterative FSI analysis.

Process for initial geometry prediction

The pattern of blade deformation depends on the properties of the materials used in manufacture and the ply stacking condition such as the main orientation, as well as the loads acting on the blade. This means that the process of predicting the initial blade geometry, which is dependent on the materials and ply stacking conditions, requires that the design geometry be determined at given propeller execution conditions.

Assuming that the optimum geometry for the propeller design condition is the "A" blade and the flexible propeller is manufactured with this geometry and is in operation, its shape would be deformed to "B". The "B" blade has completely different geometry than the optimal design in Fig. 5. Thus, the initial design process should be adapted to the geometry design of the marine propeller with the shape of the "C" blade, which is estimated by the reverse engineering method presented by Jang (2012). Then, the blade would be changed to the target geometry for the existing driving conditions. The reverse engineering algorithm for blade initial design shown below Fig. 6.

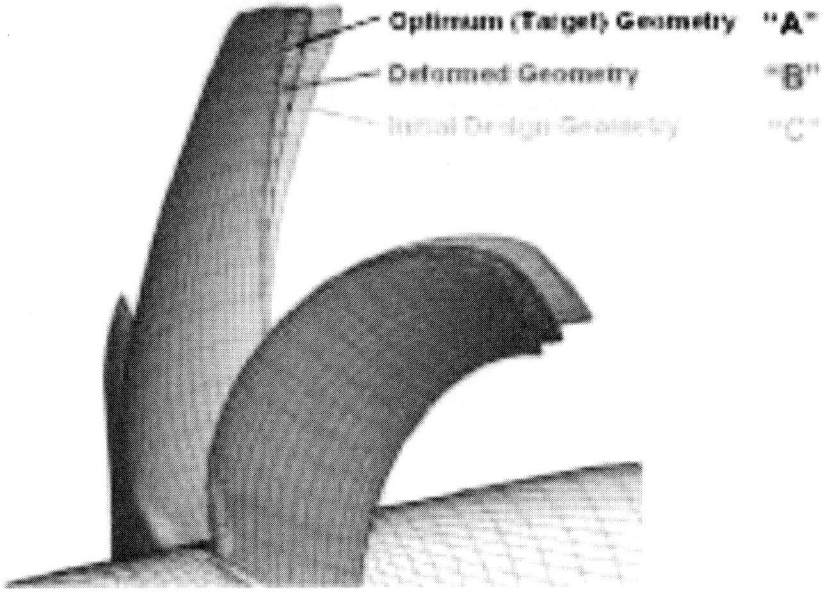

Figure 5. Blade deformation aspects of design load condition and initial design considering deformation.

Unsteady transient FSI procedure

Time domain-based FSI analysis of the propeller's blade position in the wake field is required to more accurately predict the performance and evaluate the structural strength. Rotating blades in the nonuniform inflow flow field have a "fluttering" vibration behavior caused by the change in pressure distribution and blade deformation from the response of the pressure load on the surface. Unlike the aviation propeller and wind turbine blade which act in low density air, the marine propeller cannot ignore the damping effect of the added mass on the blade surface and vibrates in the ship's wake. Thus, in this study, we considered the hydrodynamic damping effect of the added mass by applying the acoustic medium around the propeller blades.

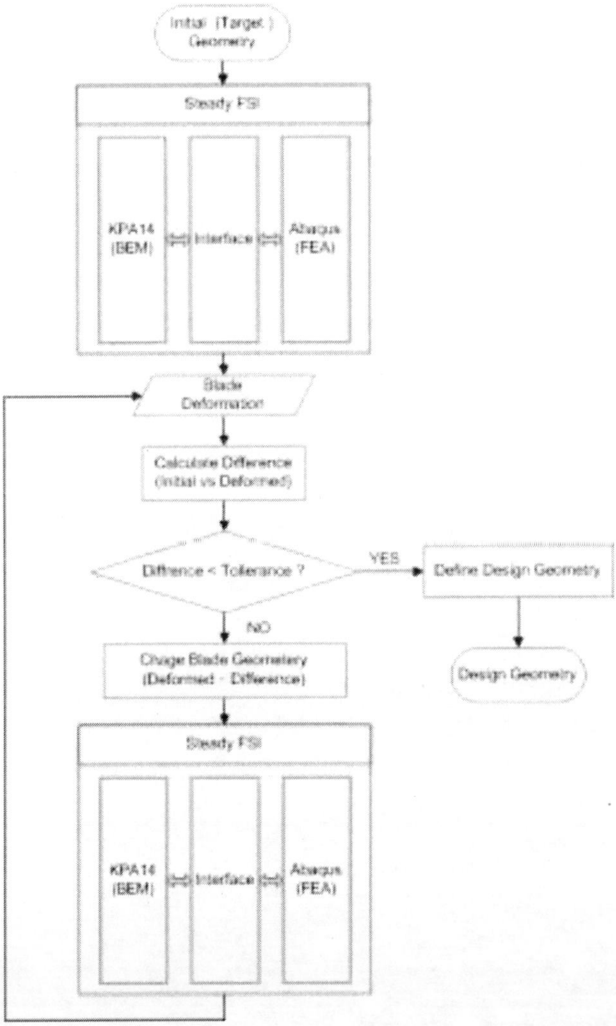

Figure 6. Reverse engineering algorithm for blade initial design.

To perform unsteady transient analysis (dynamic analysis of Abaqus Standard), it is impossible to carry out the prediction of deformed shape at $T+dt$ priory in the transient analysis. Therefore, considering the amount of infinitesimally small time step dt, and assuming pressure varies with linearly at the blade surface, the transient analysis is performed by applying blade shape of the time $T + dt$ with the shape at time T (final shape of the deformation in the previous time step) in the way that considering only changes in pressure variation of blade position in wake (Fig. 7).

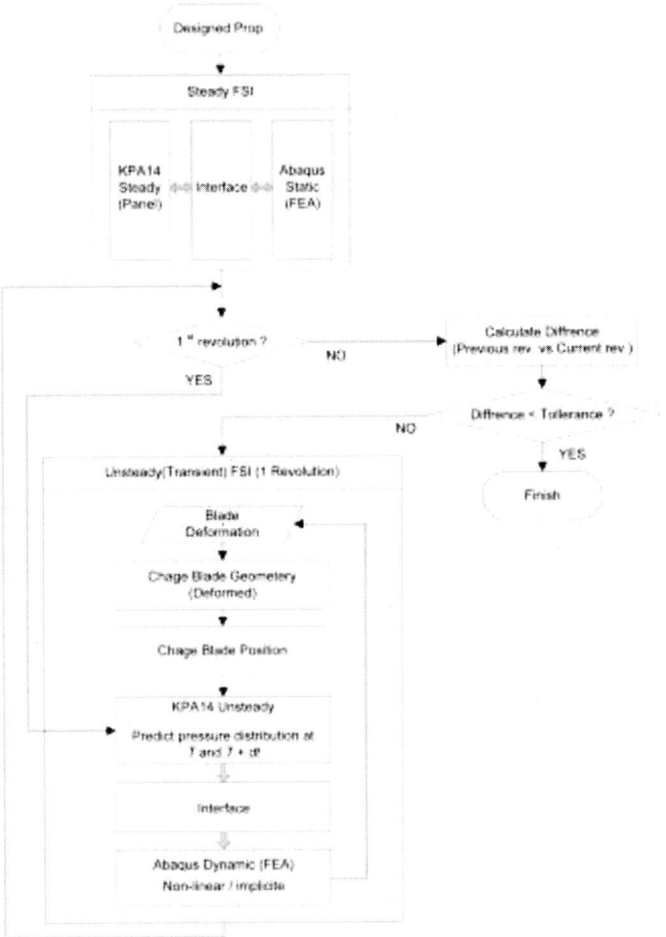

Figure 7. Transient FSI algorithm for composite propeller.

Steady analysis for the initial convergence is performed prior to the unsteady analysis in order to reduce computing time since the transient analysis can start with the result. The convergence check of the total unsteady analysis is determined by comparing the geometrical difference between the present and the former revolution positions.

VALIDATIONS

A comparison and review of experimental and computational results in the literature were conducted to verify the BEM-FEM-based steady and transient FSI analysis presented in this study. Since published references that contain both measured experimental data as well as the geometric/material properties of propellers is limited, the propellers validated in each analysis are different.

Steady state analysis

Chen et al. (2006) and Young (2008) present results for their experiments and FSI analysis of a six-bladed 610 *mm* diameter carbon fiber reinforced plastic propeller, manufactured at A.I.R. Fertigung-Technologie, GmbH in Hohen Luckow, Germany, and tested in the Naval Surface Warfare Center, Carderock Division (NSWCCD). In particular, Young (2008) conducted the validation of the proposed analytical method using the BEM-FEM FSI algorithm to compare the thrust, torque, and efficiency of a full-scaled open water model test for a "rigid" propeller P5471 and a "flexible" propeller P5479.

In this study, we validate this analytical method by comparing our results with Young's (2008) experimental and analytical results for propeller P5479. The geometry of propeller P5479 is shown in Fig. 8. This is a useful exercise for our verification because this propeller has significant hydro-elastic behavior. Isotropic homogeneous material (not a real ply stack model) is applied in the analysis. The material properties extracted by Liu et al. (2007) are $E_1 = 138.9GPa$, $E_2 = 10.2GPa$, $G_{12} = 5.7GPa$, and $\mu_{12} = 0.3$ and the main fiber orientation is applied at + 32° (see the definition of fiber orientation in Young (2008)).

The mesh size of the BEM analysis is 40× 20 for each chord (from face to back side) and span direction. We apply a cosine spacing for chord direction and a half cosine spacing for span direction to dense toward tip to faithfully reflect the leading/trailing edge and tip geometry of the blades. A total of 6400 elements (40 × 40 × 4 for chord, span, and thickness direction) are composed for the structural finite element analysis. We introduce an inverse distance weighting algorithm for the three-dimensional interpolation to share the different information at the calculating points.

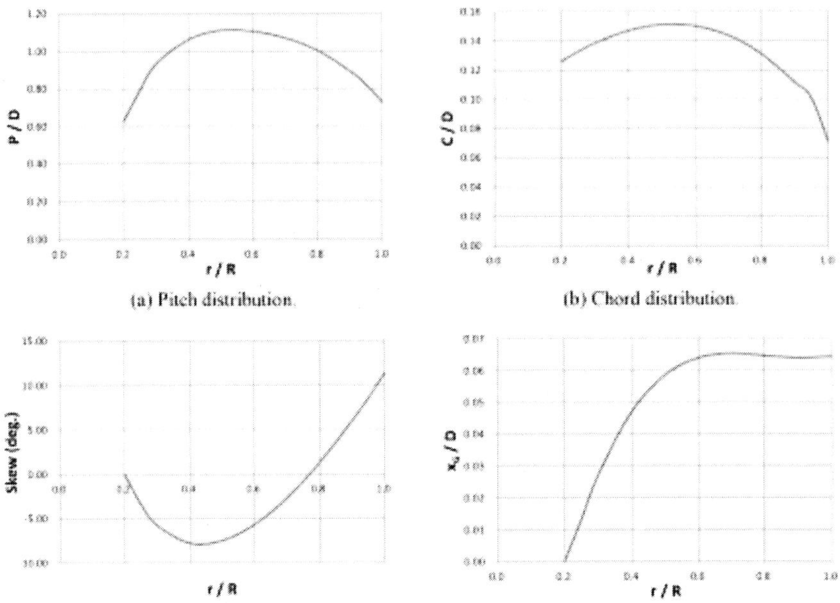

(a) Pitch distribution. (b) Chord distribution.

(c) Skew distribution. (d) Rake distribution.

Figure 8. Geometry of propeller P5479.

FSI analyses are conducted under two operating conditions, RPM = 909 and 454, and they are compared to the NSWCCD test and to Young's (2008) calculations at the advance ratio (*J*) 0.5-0.9. Fig. 9 shows that our study results and Young's (2008) analysis with the full-scale model test. It seems that the results of this study are quite good for operation advance ratio range. But, from the high advance ratio region, it contains a relatively large degree of error. It can be estimated that the reason of these errors is due to the feature of KPA14, the used BEM code, does not fully considering the viscous characteristic of Kaplan type propeller which has a finite tip chord length. Thus, it is expected to be improved by tuning the viscous correction factor in the BEM code with more tested data. The accuracy of the proposed FSI algorithm at the design advance ratio, *J* = 0.66 is sufficient for designing a composite propeller.

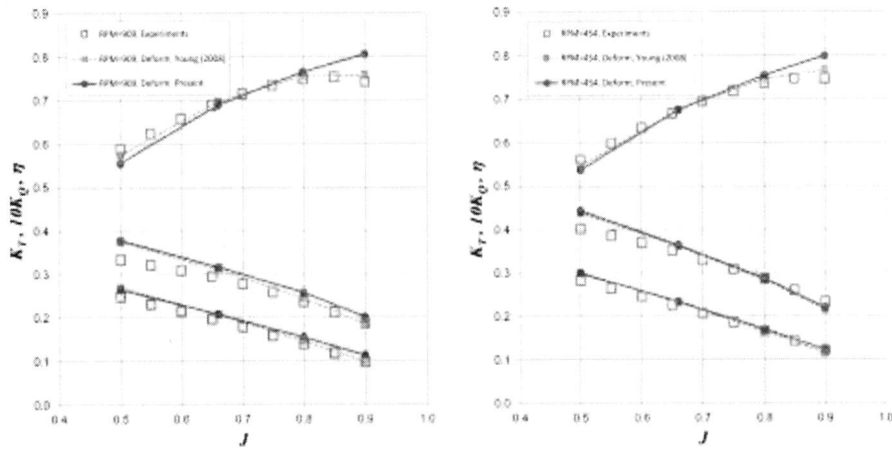

(a) RPM 909 condition. (b) RPM 454 condition.

Figure 9. Comparison of model test and BEM-FEM FSI prediction of propeller P5479.

Unsteady transient analysis

A marine propeller operates in the complex flow field condition known as the ship's wake. The composite material marine propeller, in particular, must undergo hydro-elastic response analysis for unsteady conditions because its shape changes according to the blade position in the wake. Therefore, consideration of its structural safety, such as the risk of delamination due to fatigue and accurately predicting its performance, cannot be ignored.

The damping induced by the hydrodynamic added mass significantly affects the elastic response of the propeller blade in water. Thus hydro-elastic analysis that considers the so called "wet" condition effect should be conducted. In the research of Chen et al. (2006), Young (2007a; 2008), and Motley et al. (2009), this elastic vibration effect was estimated using the potential-based BEM and the results were applied directly to the structural mass and damping coefficients.

In this study, however, the added mass effect is considered by adapting the acoustic medium field and applying it as a pressure component on the blade surface directly. We compare the results from our proposed method with the measured and calculated results of Chen et al. (2006).

The propeller used in the analysis is a P5475 (Fig. 10), which deforms in actual conditions significantly enough to validate the analysis algorithm and the Laser Doppler Velocimetry (LDV) measured result of the pitch deformation published in Chen et al. (2006). Material properties of the P5475 are $E_1 = 139.0GPa$, $E_2 = 8.06GPa$, $G_{12} = 4.88GPa$, and $\mu_{12} = 0.28$ (Young, 2007b).

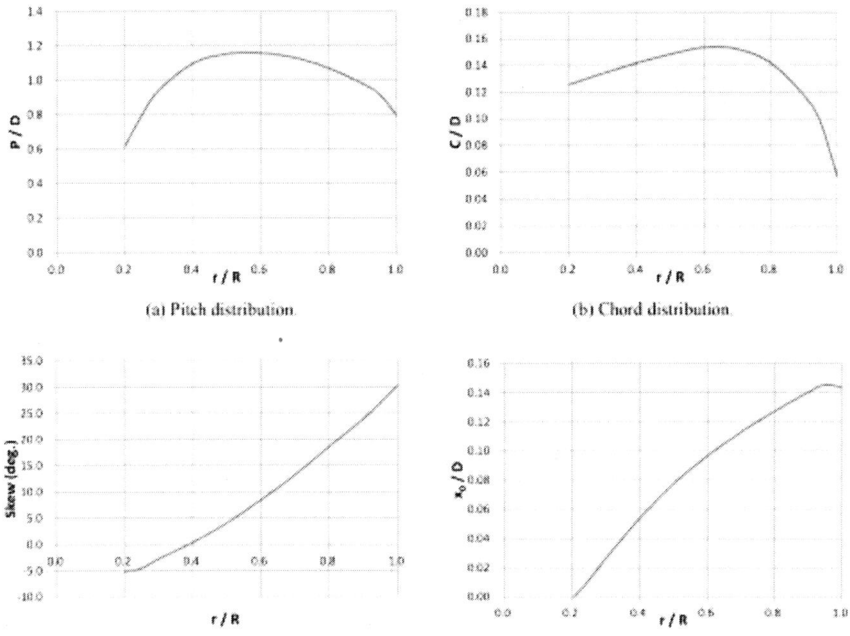

(a) Pitch distribution. (b) Chord distribution. (c) Skew distribution. (d) Rake distribution.

Fig. 10. Geometry of propeller P5475.

A wake screen was adapted to simulate the 4-cycle wake flow for measuring the hydro-elastic behavior in the unsteady flow field at NSWCCD, and the 4-cycle wake flow condition is applied to the FSI analysis. The mean velocity of inflow in the tunnel was $5.03m/s$ and velocity distribution measured by LDV is the dotted line in Fig. 11. The wake applied in this study is the solid line in Fig. 11 and it has minor differences from the measured velocity.

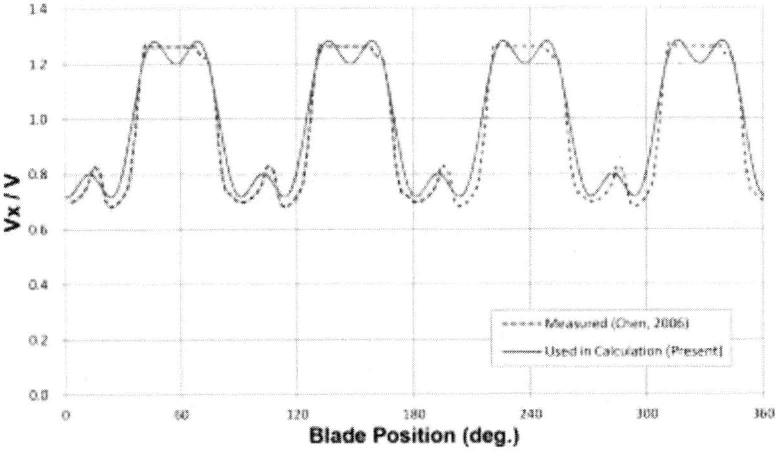

Figure 11. Inflow measured by LDV (NSWCCD) and applied to FSI analysis (present).

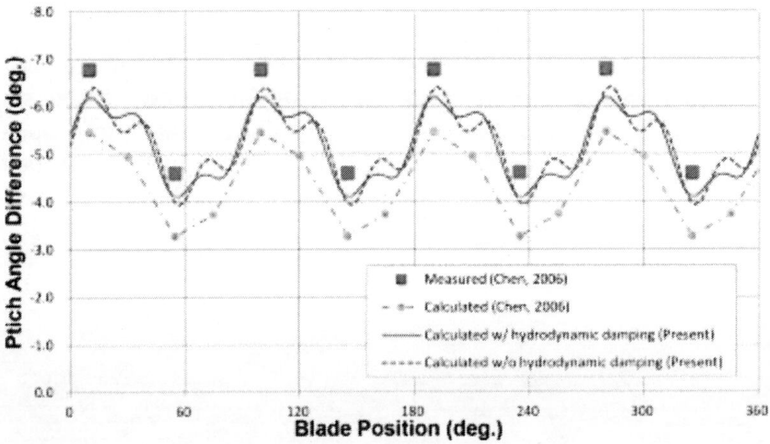

Figure 12. Pitch change for P5475 (measured, predicted by Chen et al. and present).

Pitch variations were measured by video cameras at two points, 10° and 55° . Propeller rotation was 750 *RPM* and the measured difference of pitch angle at the blade tip between the deformed and undeformed shape according blade position is represented by the squared symbol in Fig. 12. The calculation results from Chen et al. (2006) using the VLM-FEM FSI method and from this study's BEM-FEM FSI method (solid line) under the same conditions is plotted also. To examine the effect of hydrodynamic added mass damping, the "dry" condition only material damping is displayed in the dotted line.

The proposed unsteady transient algorithm is more accurate than the VLM-FEM FSI method regarding mean deformation. The difference between an analysis result that considers added mass due to the vibrating blade ("wet" condition) and a result that considers only material/structural damping ("dry" condition) is not significant. But considering that the purpose of the unsteady FSI analysis of a marine propeller in a wake field is for accurate performance prediction and structural safety evaluation, we believe that the proposed unsteady transient analysis under "wet" conditions can give more accurate and realistic prediction results for deformed geometry and variations.

The calculated variations of the thrust and torque coefficients for the key blade in the four-cycle wake flow and the measured data from the composite propeller P5475 at J = 0.66/RPM = 780 for open water conditions (Chen et al., 2006) is shown in Fig. 13. Even though differences exist in the inflow conditions, four-cycle and uniform, the average values of thrust and torque variations from the unsteady FSI analysis and the measured values from uniform flow at the same advance ratio seem very close. These results show that the proposed practical methodology is sufficiently accurate for use in composite marine propeller design with the validation results from steady state analysis.

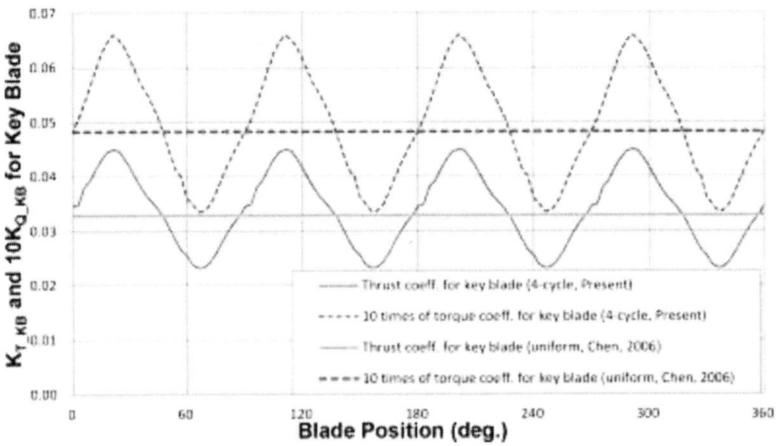

Figure 13. Thrust coefficient and torque coefficient for key blade of P5475 (present).

Natural frequency analysis for dynamic stability evaluation

Natural frequency analyses of propeller P5479 rotating at 909RPM for "dry" and "wet" conditions are conducted and the analyzed natural frequencies of each mode are compared with the measured ("wet" condition, mode 1) and calculated results of Young (2008) in Fig. 14. And the blade shapes under "wet" conditions for each mode are shown in Fig. 15.

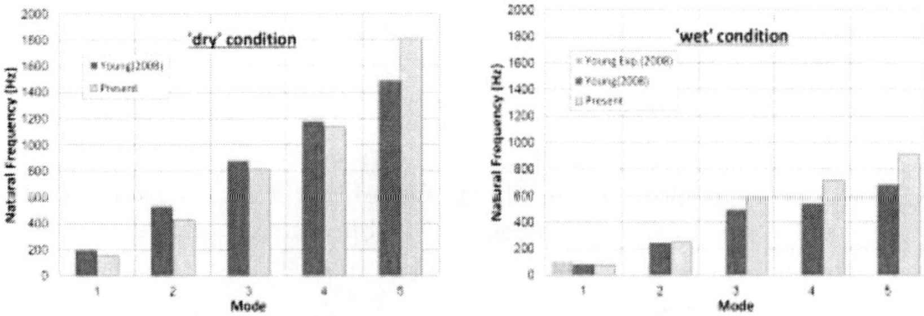

(a) 'Dry' condition. (b) 'Wet' condition.

Figure 14. Natural frequency prediction.

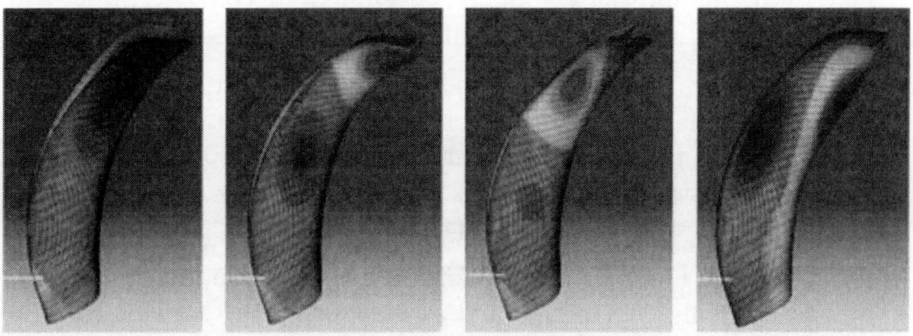

(a) Mode 1. (b) Mode 2. (c) Mode 3. (d) Mode 4.

Figure 15. Mode shapes for P5479 at 'wet' condition.

CONCLUSIONS

Since a flexible (or composite material) propeller is affected not only by the pressure changes acting on the blade but also the body forces from the rotational motion, such as the centrifugal force, the Coriolis force, and the "fluttering" hydro-elastic behavior in wake field; it is not easy to predict the propulsion performance, the initial geometry with respect to deformation, and the structural strength at the design stage.

The BEM-FEM-based practical hydro-elastic algorithm for a flexible marine propeller is developed for blade and ply orientation design to address performance prediction and structural evaluation. The study contents are as follows:

BEM code does not consider viscous effect and effective wake. However, it still gives sufficient accuracy and short computing time to apply propeller design stage. Therefore BEM-FEM FSI methodology is proposed for practical application to flexible marine propeller design instead of the more elaborate CFD-FEM FSI analysis.

1) Panel code (KPA14) is used for hydrodynamic analysis to predict pressure loads on the propeller blade and a commercial FEA tool (Abaqus 6.12 Standard, Static and Dynamic option) is used for nonlinear analysis to calculate blade deformation and evaluate strength. An interface code is developed to connect these two analysis tools and perform FSI analysis.

2) A three-dimensional continuum shell element is introduced for the FEA, which has merits in its easy modeling of composite material ply and its ability to reflect the structural behavior of objects having three-dimensional geometry.

3) The hydrodynamic damping effect of added mass, which is induced by the elastic behavior of the propeller blade, are applied to the FEM for acoustic fluid mediums and as composing system equations of motion for pressure on the surface and the structural deformation of blade geometry, unlike existing studies.

4) A validation study for the proposed steady and unsteady transient algorithm for composite propeller FSI analysis is conducted by comparing our study results with published experimental and analytical results on flexible propellers composed of orthotropic-homogeneous materials. As a result, the usefulness of the proposed practical methodology was confirmed.

Further study of the application of composite material ply structure in the FSI methodology will be conducted for further analysis of the composite material propeller. And future research will address the optimum stack algorithm that satisfies both structural strength and performance optimization for off-design operation conditions. In addition, CFD-FEM FSI methodology will be developed to improve hydro-elastic analysis accuracy.

ACKNOWLEDGEMENTS

This research was supported partly by the Industrial Convergence Strategic Technology Development Program (No. 1004-4499) funded by the Ministry of Trade, Industry and Energy (Korea).

REFERENCES

1. Blasques, J.P., Berggreen, C. and Anderson, P., 2010. Hydro-elastic analysis and optimization of a composite marine propeller. Marine Structures, 23, pp.22-38.
2. Chen, B.Y.H., Neely, S.K., Michael, T.J., Gowing, S., Szwerc, R.P., Buchler, D. and Schult R., 2006. Design, fabrication and testing of pitch-adapting (flexible) composite propellers. SNAME Propellers/Shafting Symposium, Williamsburg, VA, 9 January 2006, pp.8/1-11
3. Greeley, D.S. and Kerwin, J.E., 1982. Numerical methods for propeller design and analysis in steady flow. Transactions of Society of Naval Architects and Marine Engineers, 90, pp.415-453.
4. Hess, J.L. and Valarezo, W.O., 1985. Calculation of steady flow about propellers by means of a surface panel method. Journal of Propulsion and Power, 1(6), pp.470-476.
5. Hoshino, T., 1989. Hydrodynamic analysis of propellers in steady flow using a surface panel method. Journal of the Society of Naval Architects of Japan, 165, pp.55-70.
6. Hsin, C.Y., 1990. Development and analysis of panel method for propellers in unsteady flow. PhD thesis. MIT, Dept. of Ocean Engineering.

7. ITTC, 2005. Final report and recommendations to the 24th ITTC. The Propulsion Committee, Edinburg, 4-10 September, pp.73-136.

8. Jang, H.G., Noh, I.S., Hong, C.H. and Lee, C.S., 2012. Design algorithm of flexible propeller by fluid-structure interactive analysis. Journal of the Society of Naval Architects of Korea, 49, pp.528-533.

9. Jang, H.G., Noh, I.S., Hong, C.H. and Lee, C.S., 2013. Design algorithm of flexible propeller by fluid-structure interactive analysis. Proceeding of the 12th International Symposium on Practical Design of Ships and Other Floating Structures, Changwon City, 20-25 October 2013, pp.1158-1164.

10. Kerwin, J.E. and Lee, C.S., 1978. Prediction of steady and unsteady marine propeller performance by numerical lifting surface theory. Transactions of Society of Naval Architects and Marine Engineers, 86, pp.218-253.

11. Kerwin, J.E., Kinnas, S.A., Lee, J.T. and Shih, W.Z., 1987. A surface panel method for the hydrodynamic analysis of ducted propellers. Transactions of Society of Naval Architects and Marine Engineers, 95, pp.93-122.

12. Lee, J.T., 1987. A potential based panel method for the analysis of marine propellers in steady flow. PhD thesis. MIT, Dept. of Ocean Engineering.

13. Lee, G.H., Jang, H.G., Lee, C.S., Nho, I.S., Lee, S.G. and Hyun, B.S., 2013. Effective of manufacturing accuracy of flexible propeller on the open water performance. Journal of the Society of Naval Architects of Korea, 50, pp.349-354.

14. Lee, S.G., Byun, J.H., Paik, B.G. and Hyun, B.S., 2009. Production & performance assessment of composite material flexible propeller. Journal of the Society of Naval Architects of Korea, 46, pp.667-674.

15. Lee, S.G., Nam, J.H., Hyun, B.S., Paik, B.G., Lee, C.S., Jang, H.G. and Nho, I.S., 2012. Enhanced manufacturing and performance analysis of flexible composite propeller. Journal of the Society of Naval Architects of Korea, 49, pp.521- 527.

16. Lin, H., Lin, J., 1996. Nonlinear hydroelastic behavior of propellers using a finite element method and lifting surface

theory. Journal of Marine Science and Technology, 1(2), pp.114-124.

17. Liu, Z. and Young, Y.L., 2007. Utilization of deformation coupling in self-twisting composite propellers. Proceedings of 16th International Conference on Composite Materials, Kyoto, Japan, 8-13 July 2007.

18. Motley, M.R., Liu, Z. and Young, Y.L., 2009. Utilizing fluid-structure interactions to improve energy efficiency of composite marine propellers in spatially varying wake. Composite Structure, 90, pp.304-313.

19. Paik, K.J., 2010. Simulation of fluid-structure interaction for surface ships with linear/nonlinear deformations. PhD thesis. University of Iowa.

20. Prandtl, L., 1921. Application of modern hydrodynamics to aeronautics, National Advisory Committee for Aeronautics Annual Report 7th. USA: NASA.

21. Suh, J.C., Lee, J.T. and Suh, S.B., 1992. A bilinear source and doublet distribution over a planar panel, and its application to surface panel method. Proceedings of 19th Symposium on Naval Hydrodynamics, Seoul, 23-28 August 1992.

22. Young, Y.L., 2007a. Time-dependent hydroelastic analysis of cavitating propulsors. Journal of Fluids and Structures, 23, pp.269-295.

23. Young, Y.L. and Liu, Z., 2007b. Hydroelastic tailoring of composite naval propulsors. Proceedings of the 26th International Conference on Offshore Mechanics and Arctic Engineering, San Diego, California, USA, 10-15 June 2007, pp.777-787.

24. Young, Y.L., 2008. Fluid-structure interaction analysis of flexible composite marine propellers. Journal of Fluids and Structures, 24, pp.799-818.

3
CHAPTER

Hydrodynamic Performance Analysis of the Ducted Propeller Based on the Combination of Multi-Block Hybrid Mesh and Reynolds Stress Model

Xueming He[1*], Hecai Zhao[2], Xuedong Chen[2], Zailei Luo[2], Yannan Miao[2]

[1]*Department of Electrical and Mechanical Management, China Maritime Police Academy, Ningbo, China*

[2]*State Key Laboratory of Digital Manufacturing Equipment and Technology, Huazhong University of Science and Technology, Wuhan, China*

ABSTRACT

In order to analyze the hydrodynamic performance of the ducted propeller with high precision, this paper proposes a new method which combines Multi-Block Hybrid Mesh and Reynolds Stress Model (MBHM & RSM). The calculation errors of MBHM & RSM and standard two-equation model (standard k-ε model) on the ducted propeller JD7704 +Ka4-55 are compared. The maximum error of the total thrust coefficient K_T, the duct thrust coefficient K_{TN}, the torque coefficient K_Q and the open-water efficiency η_0 of MBHM & RSM are 2.98%, 4.01%, 1.46%, and 0.89%, respectively, which are lower than those of standard k-ε model. Indeed, the pressure distribution on the propeller surfaces, the pressure and the velocity vector distribution of the flow field are also analyzed, which are consistent with the theory. It is demonstrated that MBHM & RSM on the thruster dynamics analysis are feasible. This paper provides reference in the thruster designing of underwater robot.

Keywords: Multi-Block Hybrid Mesh, RSM, Ducted Propeller, Hydrodynamic Performance

1. INTRODUCTION

The propeller of a Remotely Operated Vehicle (ROV) is an important part, whose dynamic performance determines the motion characteristics, control performance and efficiency of underwater robot. In recent years, many scholars engage in the research of the dynamics performance of the ducted propeller. For example, Baltazar et al. [1] analyzed the open water performance of the propeller with and without duct using panel method, but the error is high. Moustafa Abdel-Maksoud et al. [2] designed a multi-component propulsor, and the hydrodynamic performance is done in both 2D and 3D with numerical method. Long Yu et al. [3] investigated the open water performance of Ka-series propeller combined with 19A duct by the panel method and the numerical method, and a 3D-geometry generate method is proposed. H. Haimov et al. [4] combined the calculation methods and model tests to calculate the hydrodynamic problem of the Ka4-60-19A, and the RANSE solution has sufficient precision.

Yet, the finite element meshes in the numeral calculation of propeller mentioned above are regular grid, and the calculation models are limited to the two-equation model, such as the standard k-ε model, which may affect the precision and speed of calculation to some extent. Considering the complexity of the geometry, flow field characteristics and the precision of calculation, a kind of multi-block hybrid mesh combined with the RSM is put forward in this paper. The convection and diffusion effect of Reynolds stress is fully considered, and the Reynolds stress is directly solved with transport equation. Finally a good result is achieved.

2. MESHING AND BOUNDARY CONDITIONS SET

Grid quality is one of the most important factors which affect the numerical precision and work efficiency in numerical calculation. Considering the complexity of the thruster structure and the advantages of structured and unstructured grids, multi-block hybrid meshing method is presented.

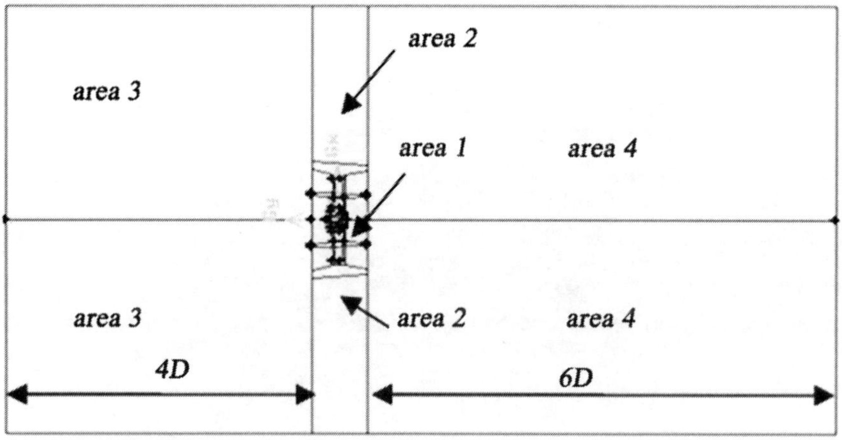

Figure 1. Division of the computational domain.

The ducted propeller is placed in a cylindrical flow field, as shown in Figure 1, the diameter of the column is 5D, the distance between the fluid entrance and front of the duct is 4D, the distance between the fluid outlet and back of the duct is 6D, the axis of the flow field coincides with the axis of the duct. The flow region of the thrust is divided into four parts. Unstructured meshes are used in the complex area adjacent to the propeller-area 1, and local crowded meshes are used near the propeller blade and hub position, and structured meshes are used in the area 2, area 3 and area 4; the grid density in area 2 is a little bigger than that in area 3 and area 4.

The grids on the surface of the propeller and the hub is generated with type of map and elements of quad, afterwards, the grids inside the duct is generated with type of TGrid and elements of Tet/Hybrid, which is shown in Figure 2. The mesh in area 2 area 3 and area 4 is generated with cooper and Hex/Wedge, which is shown in Figure 3.

Among them, TGrid and Tet/Hybrid means that the volume is divided mainly by tetrahedral cell, Hex/Wedge means that the mesh generated is hexahedral mesh, and some of them is wedge mesh, cooper means that the volume is divided according to the appointed source face.

The unstructured grid has good adaptability, reduced the difficulty in the mesh generation around the propeller, and the structured grid has good quality. The grid division method mentioned above not only reduces the grid number and improves the quality of mesh, but also improves the efficiency of the subsequent simulation.

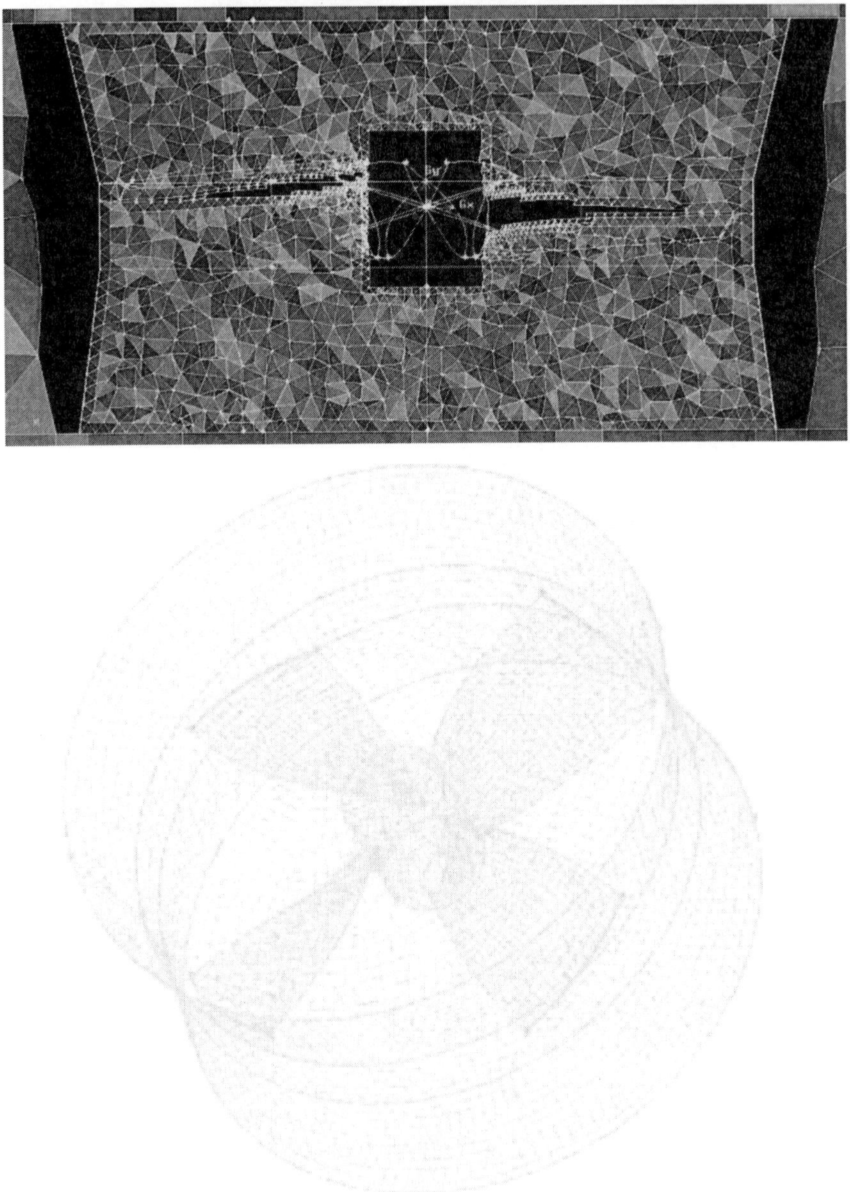

Figure 2. Grid of the interior domain.

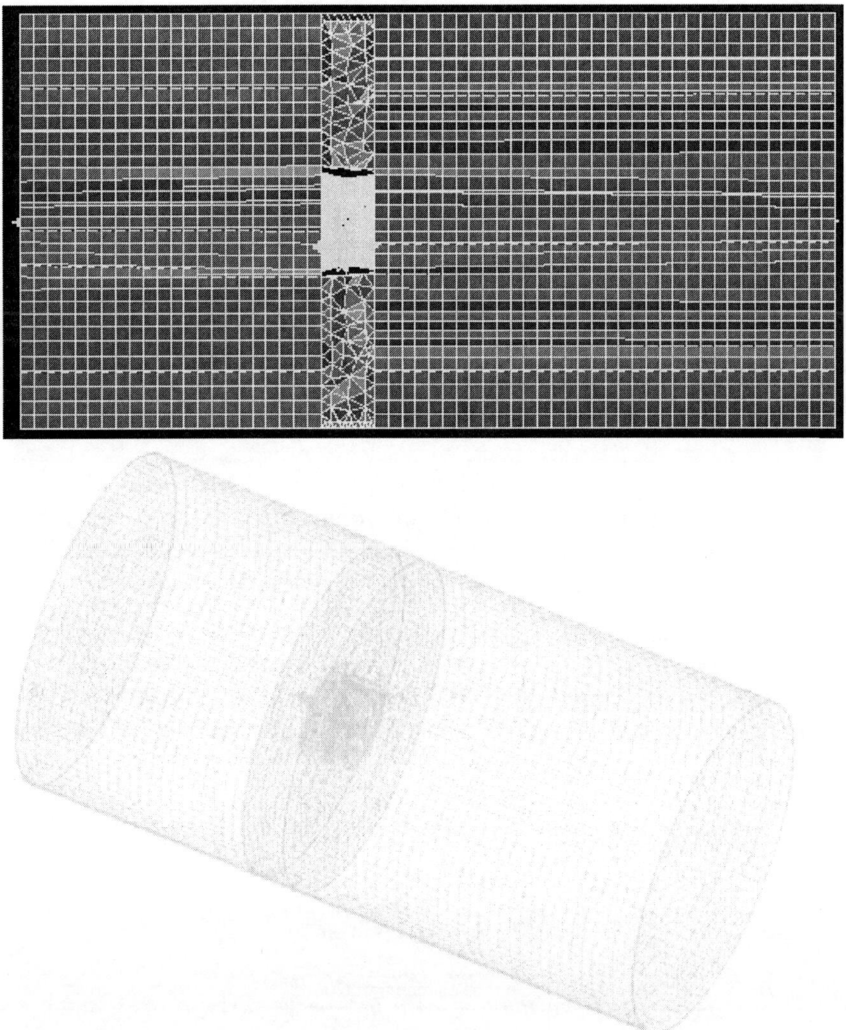

Figure 3. Grid of the exterior domain.

In general, the flow region in the boundary is almost uniform mixture in numerical calculation. For the numerical calculation only need to obtain the time averaged data of the interaction between the propeller and the catheter, so the Moving Reference Frame (MRF) model is selected. And the flow field is divided into the internal catheter flow area and the external one, the interior area is defined as a rotation domain, the external one is defined as static domain. Inlet boundary is set as velocity inlet, the magnitude is constant and the direction is perpendicular to the entrance. Outlet boundary is set as pressure-outlet; the gauge pressure is given as 0. Interfaces between the rotation domain and the static domain are set as interface. The propeller and the duct are set as solid wall with no slip conditions. Considering that the RSM is effective under fully developed turbulent condition, and turbulence around the near wall region is not stable, the non-equilibrium wall function is adopted.

3. SELECTION OF RSM MODEL AND OPEN WATER PERFORMANCE OF THE DUCTED PROPELLER

In order to use the experimental data of the JD7704 catheter +Ka4-55 propeller provided by Shanghai Jiao Tong University to verify the correct of our method and model, a matched propeller is designed. The structure is as shown in Figure 4 and its dimension in Table 1.

In the rotation of the propeller, the flow around the propeller is considered to be strong rotational flow, thus the turbulent flow is anisotropic. RSM considered the convection and diffusion of the Reynolds stress, and used the transport equation to solve the stress, avoided the viscosity hypothesis. RSM is a transport equation to solve the Reynolds stress tensor [5] , the equation is:

$$\frac{\partial}{\partial t}\left(\rho\overline{u_i u_j}\right)+\frac{\partial}{\partial x_k}\left(\rho U_k\overline{u_i u_j}\right)=-\frac{\partial}{\partial x_k}\left[\rho\overline{u_i u_j u_k}+\overline{p\left(\delta_{kj}u_i+\delta_{ik}u_j\right)}\right]$$
$$+\frac{\partial}{\partial x_k}\left(\mu\frac{\partial}{\partial x_k}\overline{u_i u_j}\right)-\rho\left(\overline{u_i u_k}\frac{\partial U_j}{\partial x_k}+\overline{u_j u_k}\frac{\partial U_i}{\partial x_k}\right)-\rho\beta\left(g_i\overline{u_j\theta}+g_j\overline{u_i\theta}\right)$$
$$+\overline{p\left(\frac{\partial u_i}{\partial x_j}+\frac{\partial u_j}{\partial x_i}\right)}-2\mu\overline{\frac{\partial u_i}{\partial x_k}\frac{\partial u_j}{\partial x_k}}-2\rho\Omega_k\left(\overline{u_j u_m}\varepsilon_{ikm}+\overline{u_i u_m}\varepsilon_{jkm}\right) \quad (1)$$

To simulate the effect of the wall to the distribution of Reynolds stress, the SSG model is chosen, which import the quadratic component of Reynolds stress anisotropic tensor to pressure and strain term. Thus the pressure and strain term is:

Figure 4. Geometry model of the ducted propeller.

Table 1. Geometry parameters of the propeller.

Parameter	Diameter(m)	Pitch ratio P/D	Disc ratio A_E/A_0	Hub diameter ratio d_0/D	Number of blades	Clearance between the blade and duct (mm)
Value	0.06	1.03	0.55	0.2	4	2

$$\varphi_{ij} = -C_1\varepsilon a_{ij} + C_2\varepsilon\left(a_{ik}a_{kj} - \frac{1}{3}a_{kl}a_{kl}\delta_{ij}\right) + C_3k\left(S_{ij} - \frac{1}{3}\langle S\rangle\delta_{ij}\right)$$
$$+ C_4k\left(a_{ik}S_{jk} - \frac{2}{3}a_{kl}a_{kl}\delta_{ij}\right) + C_5k\left(a_{ik}\omega_{jk} + a_{jk}\omega_{ik}\right) \quad (2)$$

To study the mesh-dependency of the calculation, the open water performance of the ducted propeller under 200 thousands, 300 thousands, 500 thousands grids are calculated with RSM used. Through the simulation, the propeller thrust T_P, duct thrust T_N and the total torque Q under different advance coefficient is achieved, and the total thrust coefficient K_T, the duct thrust coefficient K_{TN}, the torque coefficient K_Q and the open-water efficiency η_0 is calculated. Among them, the advance coefficient $J = \frac{V_A}{nD}$, the total thrust coefficient $K_T = T/\rho n^2 D^4$, the duct thrust coefficient $K_{TN} = T_N/\rho n^2 D^4$, the torque coefficient $K_Q = Q/\rho n^2 D^5$, the open-water efficiency $\eta_0 = \frac{K_T}{K_Q}\frac{J}{2\pi}$. In the equation, ρ means the fluid density; T means the total force generated by the ducted propeller.

The comparison of results under different grids is shown as Figure 5; we can see that the result with higher grid density is closer to the experimental result, but the results are stable in general. Considering the calculation efficiency and computational accuracy, the 300 thousands grid is chosen for later calculations.

In the calculation, the advance coefficient is defined from 0.1 to 0.8, the interval is 0.1. The simulation data and the error of the data are shown in Table 2. From the table, we can learn that the maximum error of torque coefficient K_Q is 1.46%, the maximum error of duct thrust coefficient K_{TN} is 4.01%, the maximum error of the total thrust coefficient K_T is 2.98%, and the maximum error of the open water efficiency η_0 is 0.89%. The conformity between numerical calculation and experimental data is very good, so we can conclude that the method proposed is feasible.

4. PRESSURE NEPHOGRAM ANALYSES ON THE SURFACE OF THE BLADES OF THE PROPELLER

Thrust and torque of the propeller are the two macroscopic stresses. If we want to study the force of the propel ler in water meticulously, we need to carry out the force analysis in any arbitrary point on the propeller firstly. Figure 6 shows the pressure distribution in a propeller blade surface at the design point of J = 0.394.

From Figure 6, we can learn that: 1) on the pressure surface, the pressure decreases gradually from the leading edge to the trailing edge, whereas the pressure increases gradually from root to tip. Because of the mutual interference between the catheter and blade, the leading edge of the blade and the blade tip appears pressure drop; 2) on the bake surface, the suction decreases gradually from the leading edge to the trailing edge, and yet the suction increases gradually from root to tip, and reaches the maximum in the leaf tip. There are large differences between the catheter propeller blade and ordinary propeller in pressure distribution characteristics [6] [7] .

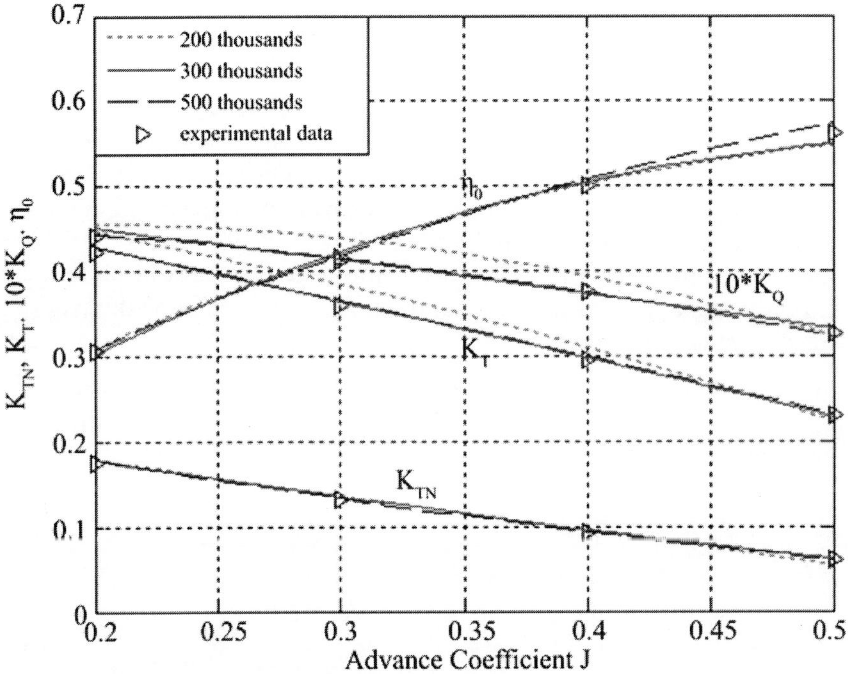

Figure 5. Grid dependency of the calculation.

Table 2. The simulation data and the error.

J	K_{TN}		K_T		$10*K_Q$		η_0	
	Simulation	Error (%)	Simulation	Error (%)	Simulation	Error (%)	Simulation	Error (%)
0.1	0.2286	0.69	0.5001	2.89	0.4576	0.47	0.1740	0.41
0.2	0.1785	1.96	0.4293	1.55	0.4443	0.96	0.3077	0.18
0.3	0.1349	3.01	0.3617	0.69	0.4137	0.11	0.4177	0.24
0.4	0.0962	3.67	0.3002	1.71	0.3747	−0.07	0.5104	0.89
0.5	0.0618	3.91	0.2351	2.11	0.3314	1.44	0.5647	0.37
0.6	0.0303	4.01	0.1688	2.98	0.2757	1.46	0.5849	0.86
0.7	0.00034	3.77	0.0958	0.26	0.2154	1.17	0.4959	−0.45
0.8	−0.0257	−9.94	0.0244	−0.56	0.1572	3.76	0.1979	−0.86

5. WORKING FLOW ANALYSIS OF THE PROPELLER

Figure 7 shows the pressure distribution of flow field of the ducted propeller. From the entrance of the flow to the blade surface of the propeller, the pressure value changes violently in the blade surface, and reaches to maximum, yet the pressure recovers on the back of the blade, and increases gradually from the back of the blade to the exit of the flow field. Thus, pressure difference exists in the blade of propeller to form thrust in the oppo site direction of the flow, which explains the working mechanism of ducted propeller.

Figure 8 shows the velocity vector of the ducted propeller at surface x = 0, we can see that the fluid velocity at both sides of the blade increases gradually from root to tip and a tip vortex forms at the blade tip.

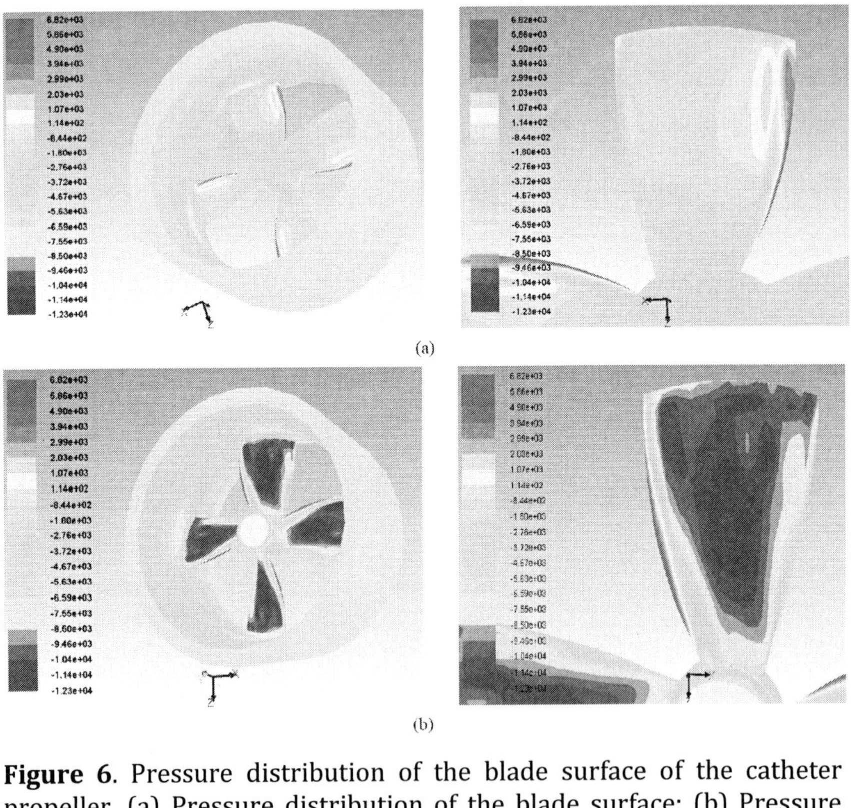

Figure 6. Pressure distribution of the blade surface of the catheter propeller. (a) Pressure distribution of the blade surface; (b) Pressure distribution of the back surface of the blade.

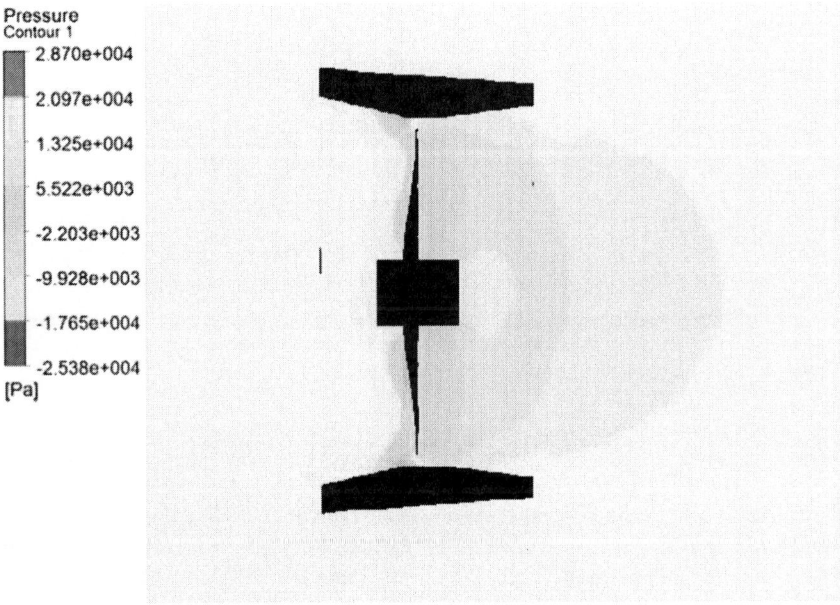

Figure 7. The pressure distribution of the flow field.

6. COMPARISON OF COMPUTATIONAL ACCURACY UNDER DIFFERENT TURBULENT MODELS

At present, the two-equation models are also used to calculate swirling flow with high Reynolds number. To verify the advantage of RSM, the standard k-ε model is used to simulate the open water performance of the ducted propeller. The open water performance calculated by RSM and standard k-ε model is shown in Figure 9; we can see that RSM has better accuracy and RSM is more applicable to calculate the open water performance of the ducted propeller.

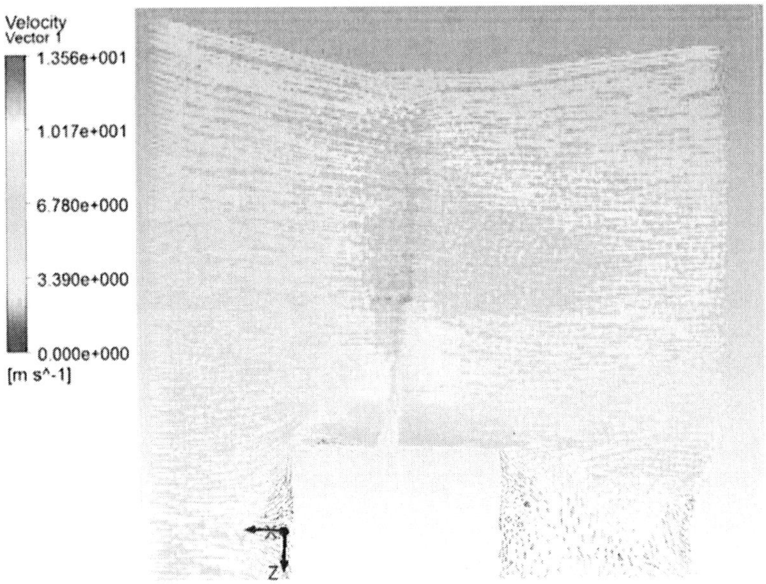

Figure 8. The velocity vector diagram.

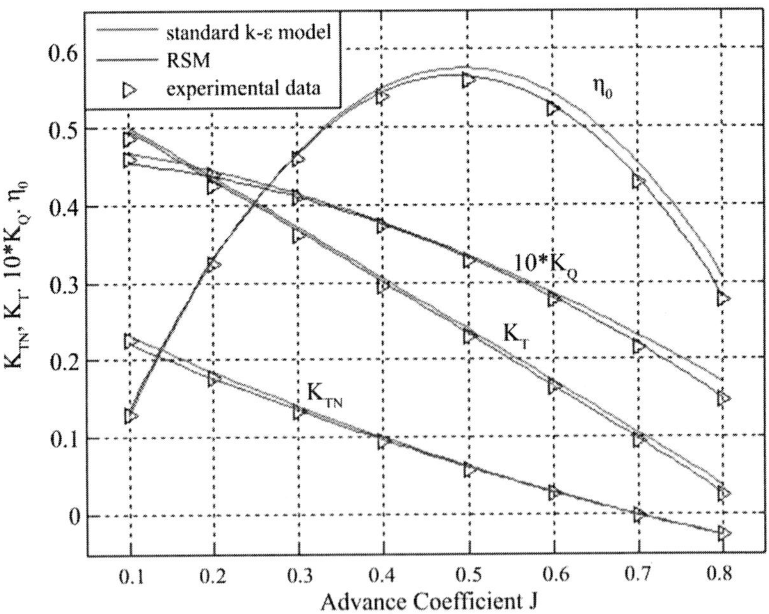

Figure 9. Open water performance calculated by RSM and standard k-ε model.

7. CONCLUSIONS

This paper proposes a new method—the combination of the multi-block hybrid mesh and RSM, then it is used to analyze the flow field of ducted propeller, including the numerical calculation on the open water performance; finally, we draw conclusions:

1) Compared with the experimental data of JD7704 catheter +Ka4-55 propeller, the calculation results based on this method are in good agreement with experimental results in the thrust coefficient, torque coefficient and open water efficiency, so the method proposed in this paper is effective and feasible.

2) The flow field distribution characteristics of the ducted propeller are analyzed in this paper; the catheter can adjust the pressure distribution on the blade surface, which reduces the pressure difference in the blade tip, but cannot completely eliminate the tip vortex.

3) In the calculation of open water performance of ducted propeller, RSM has better accuracy than standard k-ε model; the method proposed provides reference for simulation of viscous flow around the complex rotating machine.

ACKNOWLEDGEMENTS

Xueming He, Hecai Zhao, Xuedong Chen, Zailei Luo, Yannan Miao (No. 20110142130010), and the Chinese Ministry of Public Security State Research Projects (No. 2011ZDYJHJXY 012).

REFERENCES

1. Baltazar, J., Falcão De Campos, J. and Bosschers, J. (2012) Open-Water Thrust and Torque Predictions of a Ducted Propeller System with a Panel Method. International Journal of Rotating Machinery, Article ID: 474785.

2. Abdel-Maksoud, M., Steden, M. and Hundemer, J. (2010) Design of a Multi-Component Propulsor. Proceedings of 28th

Symposium on Naval Hydrodynamics, Pasadena, 12-17 September 2010, 744-758.

3. Yu, L., et al. (2013) Numerical Analysis of Ducted Propeller Performance under Open Water Test Condition. Journal of Marine Science and Technology, 18, 381-394.

4. Haimov, H., et al. (2010) Ducted Propellers. A Solution for Better Propulsion of Ships Calculations and Practice. Proceedings of 1st International Symposium on Fishing Vessel Energy Efficiency E-Fishing, Vigo, 265-271.

5. Peng, H.H., Qiu, W. and Ni, S. (2013) Effect of Turbulence Models on RANS Computation of Propeller Vortex Flow. Ocean Engineering, 72, 304-317.

6. Pan, X., Zhang, J.-Y. and Liu, C.-J. (2011) Numerical Simulation and Analysis of Cavitation Characteristics of Ship Propeller. Intelligence Information Processing and Trusted Computing (IPTC), Wuhan, 22-23 October 2011, 180-183.

7. Choong, W.H., et al. (2013) Outboard Marine Propeller Performance Analysis through CFD Modelling. Computer Modelling and Simulation (UKSim), Cambridge, 10-12 April 2013, 310-323.

4

CHAPTER

Computational Development of Marine Propeller Design

Kiam Beng Yeo, Wen Yee Hau and Cheah Meng Ong

Unit kajian Bahan dan Mineral ,Sekolah Kejurteraan dan Teknoogi Maklumat, Universiti Malaysia Sabah, jalan UMS,88400 Kota Kinabaluy, Sabah, Malaysia

ABSTRACT

This study demonstrates the computational development of marine propeller, starting with the formation of the three-dimensional model through computational software, followed by the fluid dynamics analysis for different profiles such as number of blades, pitch to diameter ratio, rake angle and skew angle. Analysis results showed that marine propeller with three or four blades have better performance while increasing of skew angle decreases the swirling problem. The importance of mesh refinement and locally defined cell sizes are also presented.

INTRODUCTION

Marine propeller design is limited by constraints of hydrodynamic parameters such as Reynolds number, cavitations phenomenon and maximum diameter for weight loading (Justin, 1986). Since water are denser compared to air in **physical properties**, so the ability of lift force production for thrust amount on each unit of the blade area are also limited. A good computational method to develop and investigate the flow dynamics of marine propeller is important for three-dimensional flow analysis. Theoretical calculation is too complex to consider the irrotational non-uniform fluid behaviour on the inflow of propeller back surface, mixing of irrotational and rotational flow on propeller revolving region and rotational flow behind the face surface of marine propeller.

Rake, skew and pitch angles are some critical parameters in marine propeller analysis. Small rake can reduce drag force on propeller surface, hold water better for each rotation, thus increasing the thrust force and efficiency of the propulsion system. Ghassemi (2009) 5 skew angle profiles found that skew formation on the propeller design could provide stability but low skew angle would induce vibration to the propulsion system.

Pitch angle is the changes of helix angle of the complete blade section located at the root section of the propeller blade. Increasing of pitch to diameter ratio, p/D reduce the interaction area with inflow fluid but increase the face interaction to the rotational motion as it faces perpendicular to the acting rotational force with fluid (Bernitas and Ray, 1982). Theoretically, larger blade area increases the sweeping area perpendicular to the water surface as it rotates which results in higher torque (Oosterveld, 1975). However, increasing of blade area also reduces the efficiency of peak performance curve at the same time (Tsakonas *et al.*, 1972).

Computational Fluid Dynamics (CFD) is a very useful approach used routinely in optimization and design for assessment of marine engineering system (Mihaela, 2005). It uses Navier-Stoke equation to solve the nonlinear flow for marine propeller especially when dealing with turbulent, shock wave and break wave (Miyata, 1997). The software decomposes spatial domain into tiny cells to form the volume mesh and a suitable algorithm is applied to solve the equation of motion (Wu, 2010).

Subhas *et al.* (2012) conducted computational estimation of thrust and torque performance of marine propeller for various rotational speeds found an extremely ideal or same result compared with experimental work. In Liu *et al.* (2012), the uncertainties in CFD had lowered the thrust and torque performance by 5% to be the cause of software code lacking of grid density or too much of numerical dissipation in the vicinity of cavity-fluid interface. Researchers such as Takayuki *et al.* (2003) and Shin and Shitalkumar (2003) have also shown successful analysis work of marine propeller using CFD.

GENERATION OF MARINE PROPELLER BY CAD

Initially, the Computer Aided Design (CAD) solid model of the propeller is developed via a two-dimensional coordinate database of the specific hydrofoil series. Two major types of pitch distribution, commonly offered and manufactured, are the full distribution and 80% pitch distribution. Full distribution has a pitch distance as from the hub to tip while 80% pitch distribution is formed by taking 80% of pitch length from the hub section and slowly increased until the complete distribution as the design requirement. Figure 1 shows the pitch distribution of two common profiles of marine propeller design.

Gerr (2001) suggested that if the blade angle from hub to tip is the same it will produce tip flow that race ahead of the root section, as the tip rotates with faster linear velocity in a single rotational.

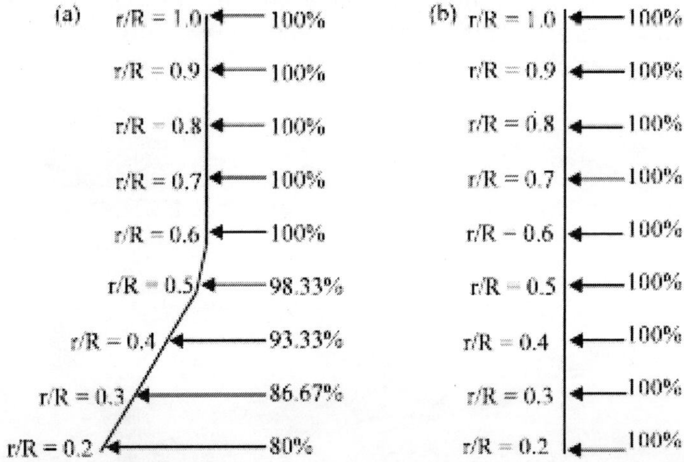

Figure 1(a-b): (a) An 80% pitch distribution and (b) Full distribution of marine propeller

From the hydrofoil coordinate database, size of hydrofoil shall be examined due to different enlargement required for the diameter of the propeller for the specific size of blade section. Also, the hydrofoil that formed shall have various attack angles that is needed to be adjusted by law of cosines for the coordinate system. This angle of attack for each hydrofoil is wrapped on cylindrical surface that has various sectional radials from two-dimensional plain surfaces to three-dimensional complex surfaces. Three-dimensional hydrofoils are drawn by 3D sketches, loop by surface loop function, then formed to solid body from surface body. By using the required hub size and length, required number of blades is added evenly around the circular hub. Figure 2 displayed the process of the formation for the marine propeller by using the CAD technique.

FLOW TRAJECTORIES WITH INFLOW, PROCESSING AND OUTFLOW

In the comparison of flow trajectories that consisted of inflow and outflow for marine propeller, 2 blades propeller does not have swirling of fluid on the face surface of the hub. Then, the swirling condition increased as the blades number increased. Low passed of inflow and high cutting capacity of blades provides more possibility for propeller to cause this vortex phenomenon to occur.

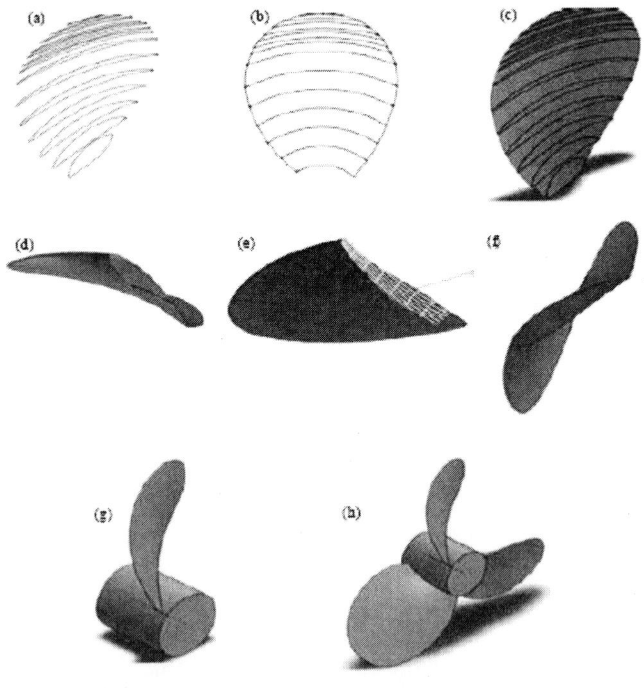

Figure 2(a-h). Three-dimensional marine propeller generation by CAD software

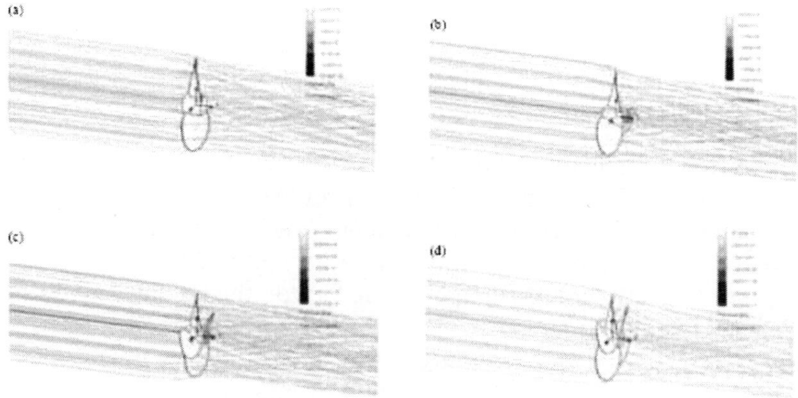

Figure 3(a-d). Flow trajectories for marine propeller with different No. of blade, (a) 2, (b) 3, (c) 4 and (d) 5 blades

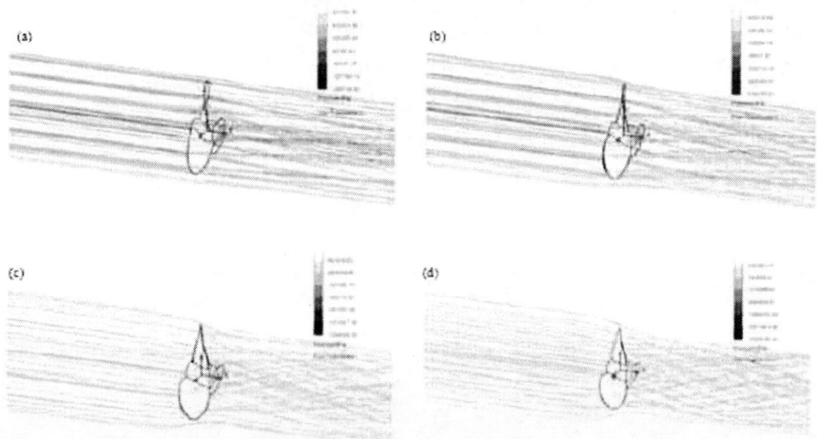

Figure. 4(a-d) .Flow trajectories of 3 blades marine propeller at different p/D profile, (a) p/D = 0.6, (b) p/D = 0.8, (c) p/D = 1.2 and (d) p/D = 1.4

Also, as the blade number increased to 5, a straight and concentrate flow that has higher flow pass, remain as small cylindrical flow on the centre of the face hub. From the analysis, 3 and 4 blades marine propeller would be the best selection, where 3 blades offer good performance, acceleration and top end and 4 blades come with better hole shot, stability and fuel economy. Figure 3 shows the flow trajectories for marine propellers with different number of blades at p/D = 1.0, 0° rake angle and 0° skew angle.

Computational results show that the pitch to diameter ratio, p/D has a concentration of flow vortex on low p/D profile that is closed to the face of the hub surface, where the rotational outflow becomes highly focused and more stable on its flow pattern. However, p/D ratio doesn't have a consistent formation of swirling near the face on the hub at the outflow. Higher p/D profile does not come with vortex phenomenon as other types of propeller geometry. Figure 4 also shows the flow trajectories of 3 blades marine propeller at different p/D profile.

Rake angle profile does not come with clear trend of flow trajectories compared to other geometric changes. From the Fig. 5 it can be observed that 0 and 20° of rake angle had high swirling flow phenomenon occurring on the hub surface with 20° comes slightly small compared to 0°. While, 10 and 30° rake angle does not face any of the vortex flow problem. Since mixed flow does not occur on these profiles, the outflows of these two geometries were stable with less loss.

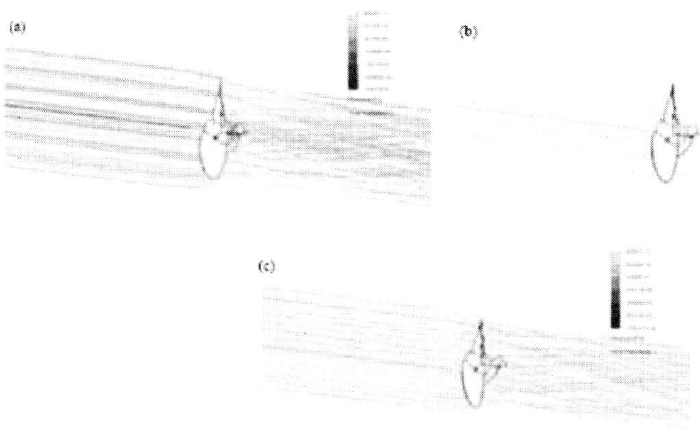

Figure 5(a-c). Flow trajectories of 3 blades marine propeller at different rake angle, (a) 10° rake angle (b) 20° rake angle and (c) 13° rake angle

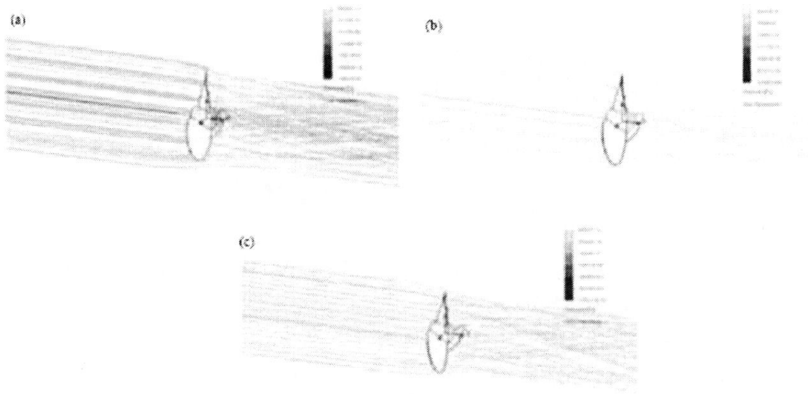

Figure 6(a-c). Flow trajectories of a 3 blades marine propeller at different skew angle (a) 10° skew angle (b) 20° rake angle and (c) 13°rake angle

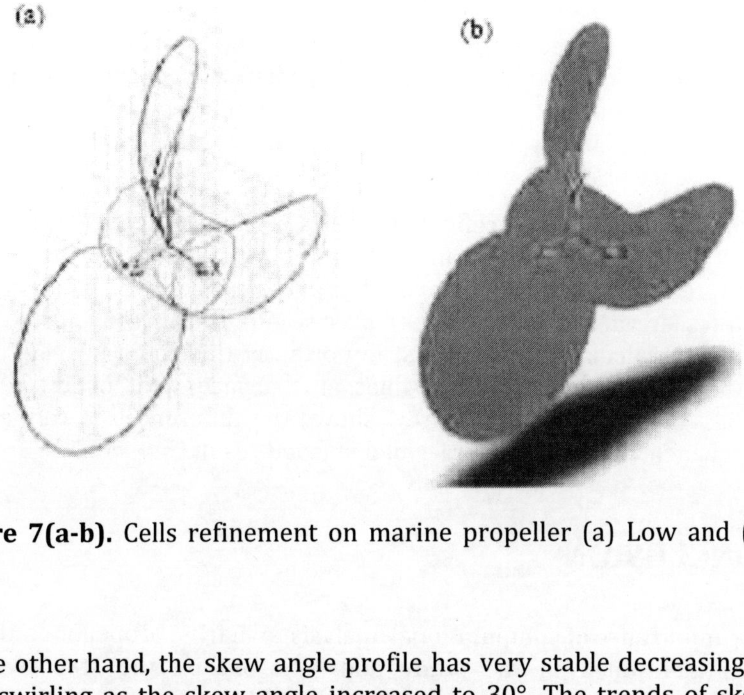

Figure 7(a-b). Cells refinement on marine propeller (a) Low and (b) High

In the other hand, the skew angle profile has very stable decreasing of flow swirling as the skew angle increased to 30°. The trends of skew angles were stable and do not face any swirling of flow that formed vortex phenomenon as other profile. Thus, this propeller geometry has high efficiency because it does not have any losses of flow that dies off itself near to face of the hub surface as others.Figure 6 also shows the flow trajectories of a 3 blades marine propeller at different skew angle.

MESH GENERATION AND REFINEMENT

The refinement of numerical data from CFD analysis could increase the accuracy of the finding. Since flow simulation in CFD analysis software is used to generate the meshes it is split with various x, y and z planes that orthogonal to Cartesian global coordinate system by half each. For the marine propeller, the load distribution over the propeller surface is concerned and should have more a precise and smaller size of cells distribution for a complete range of studies.

In the other hand, the fluid medium surrounding the propeller has less concentration, the cells size can be larger compared to the propeller does not make much differences for in and out flow which is far from the propeller body. Enhancement of cells requires a much more precision on the edge and tip section of the propeller, since these regions are of much thinner sectional area with high changes of angle. In normal procedure of mesh refinement without stressing on the edge it would cause improper distribution and iteration that focus on unnecessary or less important regions.

Higher level of meshes refinement certainly would decrease the error but it will consume large amount of computational time instead. An accurate way of refinement would provide a sum of calculation data with lesser variances of error or precise numerical result for a set of repeating calculation. Also, most importantly this could provide a fine plotting of curve lines as iso-lines and surfaces plot over the solid surface of the propeller. Figure 7 shows the differences of refinement level which affects the accuracy of displayed result.

CONCLUSION

The importance of computational analysis to marine propeller study has been mentioned in the study and followed by the step by step generation of marine propeller model in CFD software. CFD analysis shows that 3 and 4 blades marine propeller would be the best selection, as it offers good performance, stability and fuel economy as compared to the rest. Computational results also showed that pitch to diameter ratio, p/D does not have a consistent formation of swirling near face on hub at outflow. Besides, the skew angle profile has very stable decreasing of flow swirling as the skew angle increased to 30° but the rake angle profile does not come with proper trend of flow trajectories compared to other geometric changes.

For propeller blade, optimization is highly needed on the edge or tip of the blade. These regions consisted of curvature with tip section that formed from hydrofoil section has sharp trailing edge. The smaller cells which enhanced or improved produces much rapid calculation for either fluid to solid, fluid to fluid and solid to solid cells interaction allows increases of accuracy in CFD analysis.

ACKNOWLEDGMENT

The authors would like to express their appreciation to the Unit Kajan Bahan dan Mineral (Materials and Minerals Research Unit), School of Engineering and Information Technology, Universiti Malaysia Sabah and the Ministry of Higher Education of Malaysia for their support.

REFERENCES

1. Bernitas, M. and D. Ray, 1982. Optimal revolution B-series propellers. Department of Naval Architecture and Marine Engineering, Michigan, USA. http://deepblue.lib.umich.edu/handle/2027.42/91757.
2. Gerr, D., 2001. The Propeller Handbook: The Complete Reference for Choosing, Installing and Understanding Boat Propeller. McGraw-Hill, New York, ISBN: 9780071381765, Pages: 176.
3. Ghassemi, H., 2009. The effect of wake flow and skew angle on the ship propeller performance. Trans. B. Mech. Eng., 16: 129-158.
4. Justin, E., 1986. Marine propellers. Annu. Rev. Fluid Mech., 18: 367-403.

5. Liu, D., F. Hong, F. Zhao and Z. Zhang, 2012. The CFD analysis of propeller sheet cavitation. China Ship Scientific Research Center, Wuxi, China.

6. Mihaela, A., 2005. Developments in the design of ship propeller. Scientific Bulletin of the Politehnica University of Timisoara Transactions on Mechanics, University of Galati.

7. Miyata, H., 1997. Time-Marching CFD Simulation for Moving Boundary Problems. In: Twenty-First Symposium on Naval Hydrodynamics, NSB and NRC (Eds.). National Academy, USA.

8. Oosterveld, M.W.C., 1975. Further Computer-Analyzed Data of the Wageningen B-Screw Series. MSMB. Publisher, Rotterdam, Pages: 14.

9. Shin, H.R. and J. Shitalkumar, 2003. CFD validation for a marine propeller using an unstructured mesh based RANS method. Proceedings of the 4th ASME/JSME Joint Fluids Engineering Conference, July 6-10, 2003, Honolulu, Hawaii, USA -.

10. Subhas, S., V.F. Saji, S. Ramakrishna and H.N. Das, 2012. CFD analysis of propeller flow and cavitation. Int. J. Comput. Applic., 55: 26-33. |

11. Takayuki, W., K. Takafumi, T. Yoshihisa, M. Masatsugu and S.H. Rhee, 2003. Simulation of steady and unsteady cavitation on a marine propeller using a RANS CFD code. Proceedings of the 5th International Symposium on Cavitation, November 1-4, 2003, Osaka, Japan -.

12. Tsakonas, S., W.R. Jacobs and M.R. Ali, 1972. An exact linear lifting-surface theory for a marine propeller in a nonuniform flow field. Report SIT-DL-72-1509, Davidson Laboratory, New Jersey, USA., February 1972.

13. Wu, X., 2010. A Rapid Development Process for Marine Propellers through Design, Simulation and Prototyping. Memorial University of Newfoundland, St. John's, Newfoundland and Labrador, Pages: 244.

5

CHAPTER

Investigation of Different Methods of Noise Reduction for Submerged Marine Propellers and Their Classification

Mohammad A. Feizi Chekab[1], Parviz Ghadimi[1,], Seyed Reza Djeddi[1], Mehdi Soroushan[1]

[1]Department of Marine Technology, Amirkabir University of Technology, Tehran, Iran

ABSTRACT

In this paper, the most prevalent methods of reducing noise from the marine propellers are surveyed and introduced. Subsequently, these methods are analyzed and classified as five different categories. This categorization is conducted from the stand points of technology and cost, the type of noise (cavitation, non-cavitation noise), the method of noise reduction (change in propeller geometry, modification of inlet flow, and confinement of the propeller), frequency of the targeted noise, and simplicity of implementation of the method of noise reduction in the existing vessels. The scientific classification performed in this article would bring about better recognition of and acquaintance with each of the mentioned methods of noise reduction, its limitations in cost and applications, which would in turn help the designers and the decision makers make the right decision under different operational circumstances.

Keywords: submerged marine propeller, sound propagation, methods of noise reduction, cavitation noise

1. INTRODUCTION

One of the most important parameters in designing the marine propulsion systems is the acoustic noise due to the oscillations of pressure in the water surrounding the propeller. Consequently, a lot of efforts have been made to minimize the acoustic noise generated by the propulsion system which accounts for the major source of noise propagation from vessels [1]. In addition to the acoustic pollution that this propagated noise can cause in marine environment [2, 3, 4, 5, 6], it can also reduce the pleasure and comfort of marine transportation. On the other hand, propagated noise from the commercial vessels may interfere with the natural echolocation or biosonar systems of whales and dolphins which they use to locate predators and preys and this can confuse them and interfere with their basic biological functions such as feeding and mating [3]. The close resemblance between the propagated noise signals of some vessels and the noise signals of particular species of whales has caused confusion for these mammals in recognizing the ships and has endangered both the whales and the vessels [4]. Last but not least, one of the major methods used in the identification of naval ships and submarines is based on their radiated noise signals.

In the present paper, main tools of propeller noise reduction are presented due to the fact that propellers are largely used as the main propulsion system for major commercial, passenger and naval ships. These methods are investigated and classified from different points of view. These classifications will pave the way for the designers and manufacturers in choosing the best suitable method for their particular operational circumstance.

The history and background of all the reduction tools and the scholars who have done prime research on the development of these methods are presented in the respective sections of each approach. Therefore, the background of the aforementioned methods is not given in the introduction.

There seems to be multitude of tools and ways for reducing the propeller noise which are presented one after another and finally are given inside a general classification that is suggested by the writers of the present work.

2. BLADE GEOMETRY MODIFICATION

Reducing pressure oscillations on the propeller is one of the most effective ways of reducing the radiated noise. Different parameters will affect these oscillations, as follows:

1. Increase in Skew Angle
2. Increase in number of blades
3. Increase in blade area
4. Suitable distribution of blade pitch
5. Trailing edge geometrical modifications
6. Propeller blade finishing fineness

Before moving to next section, it is necessary to explain the geometrical properties and parameters of a conventional marine propeller. Figure 1 shows a general definition of a propeller's blade section.

One of the most important parameters that defines the propeller's geometry is Pitch. This parameter refers to the helical progress along a cylindrical surface [7]. This helical progress is presented in Figure 2.

Another factor which is often referred to especially in the case of low-noise propellers is the Skew Angle. This parameter which is defined and shown in Figure 3, is the angle between a line that is drawn through the hub center and a point located in the mid-chord position of a particular section and the directional line.

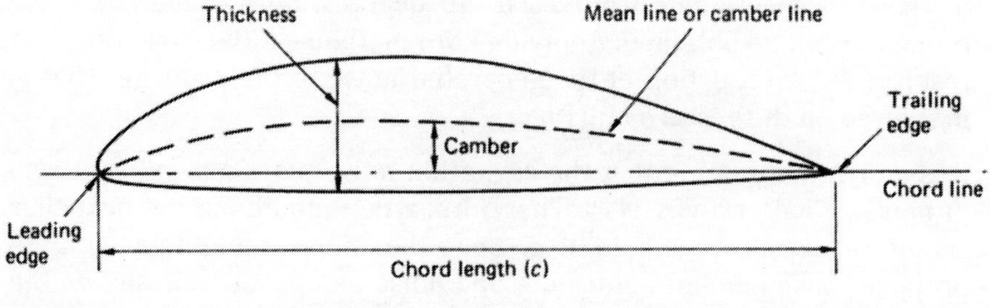

Figure 1. General definition of a propeller's blade section [7]

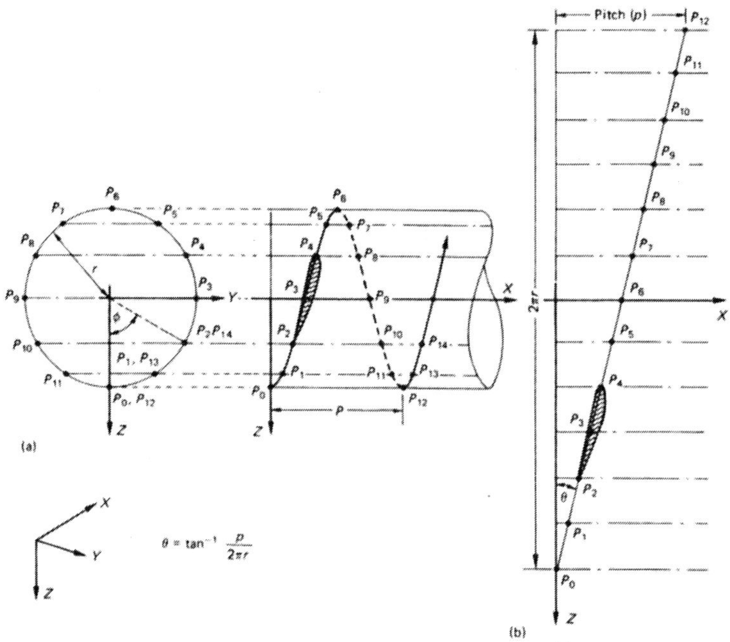

Figure 2. Definition of pitch: (a) helix definition on a cylinder of radius r and (b) development of this helix on the cylinder [7]

After testing and investigating nine standard propellers in the operational conditions using a 250 hp engine that turns the propeller with the rotational velocity of 1500rpm, it has been shown that an increase in the number of blades from 3 to 5 can lead to approximately 55% reduction in the propagated noise of the propeller. Also with an increase in skew angle from 12 to 40 degrees, a 48% average noise reduction can be obtained. Applying both methods at the same time can result in 77% reduction of the generated noise of the propeller that is measured on the vessel's hull [8].

It is worth mentioning that the cavitation noise plays a significant role in noise signal of a vessel [9]. Therefore, when modifying the propeller geometry, extra care should be given to the effects it might have on the propeller based on both efficiency and noise perspectives. Based on the above, Koushan et al. [10]investigated a systematic method for the optimization of the propeller blades from cavitating and acoustical points of view.

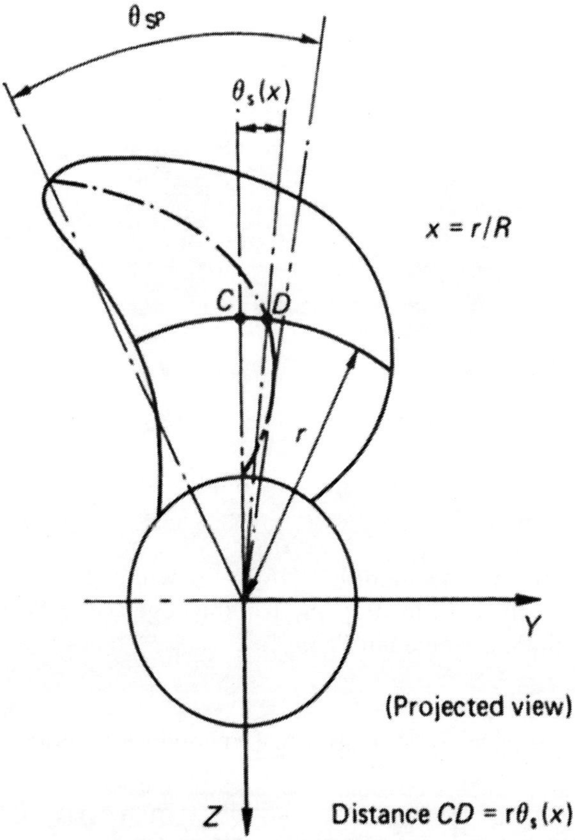

Figure 3. Definition of the propeller's Skew angle [7]

Another experiment in this field was done by Andersen et al. [11] on a Kappel-Type propeller. In this study, two 7 and 8-blades propellers were investigated. It was shown that, with the increase of propeller blades from 7 to 8, 15dB noise level and 80% thrust oscillations reduction can be achieved. The results of these experiments are given in Figure 4.

Other experimental studies were done by Sharma [12] in India on propeller noise. He used three different types of propellers, i.e. A, B and C.

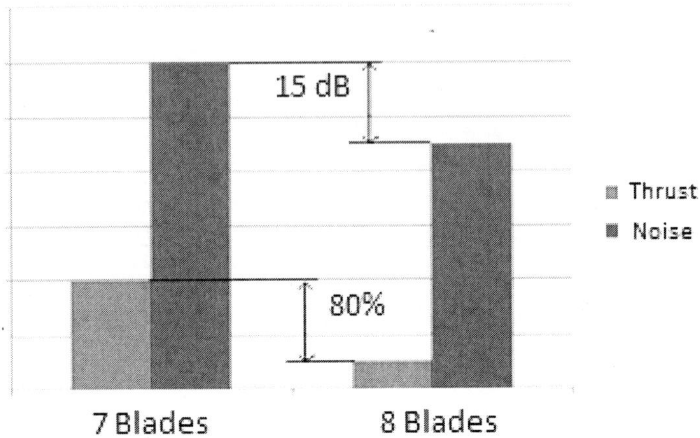

Figure 4. Results of the experiments on Kappel-propellers

Propeller A was a TMB-modified NACA-66 while propeller B was a Plano-convex with sharp leading and trailing edges. Specifications of the propellers A and B are given in Table 1.

Table 1. Geometrical Specifications of Propellers A and B

r/r_0	C_m/c	t_m/c	c/D	P/D
0.25	0.037	0.253	0.174	1.092
0.35	0.0388	0.192	0.193	1.093
0.45	0.0372	0.146	0.212	1.092
0.55	0.034	0.113	0.227	1.09
0.65	0.0312	0.087	0.232	1.087
0.75	0.029	0.068	0.226	1.085
0.85	0.027	0.052	0.202	1.084
0.95	0.0247	0.045	0.139	1.083

Propeller C was a Wageningen B4-55 with the specifications given in Table 2.

Table 2. Geometrical Specifications of Propeller C

r/r_0	t_m/c	c/D	P/D
0.3	0.117	0.247	0.881
0.4	0.101	0.273	0.934
0.5	0.084	0.29	0.982
0.6	0.068	0.296	0.992
0.7	0.057	0.289	1.004
0.8	0.048	0.265	0.995
0.9	0.043	0.204	1.005

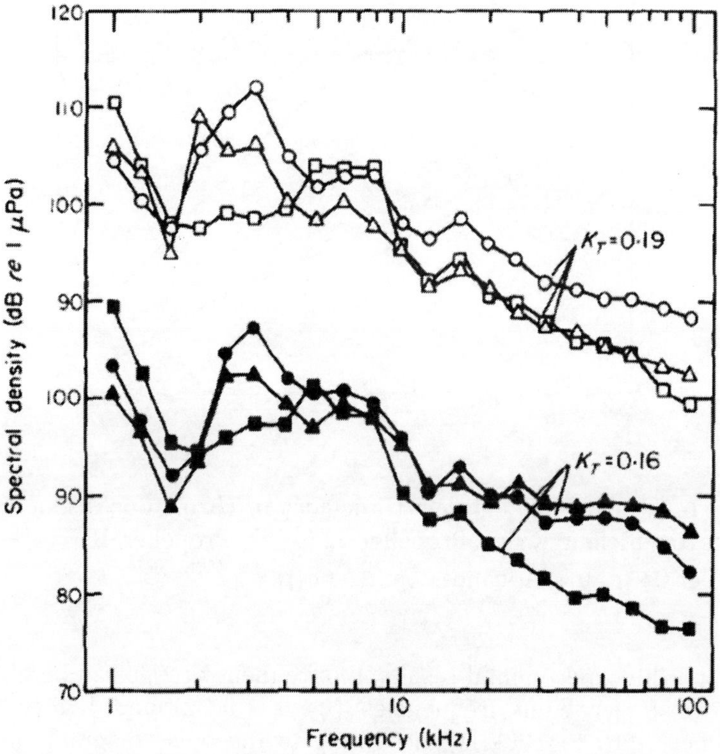

Figure 5. Propeller noise in two different operational conditions: Propeller A (circle), Propeller B (triangle),Propeller C (rectangle)

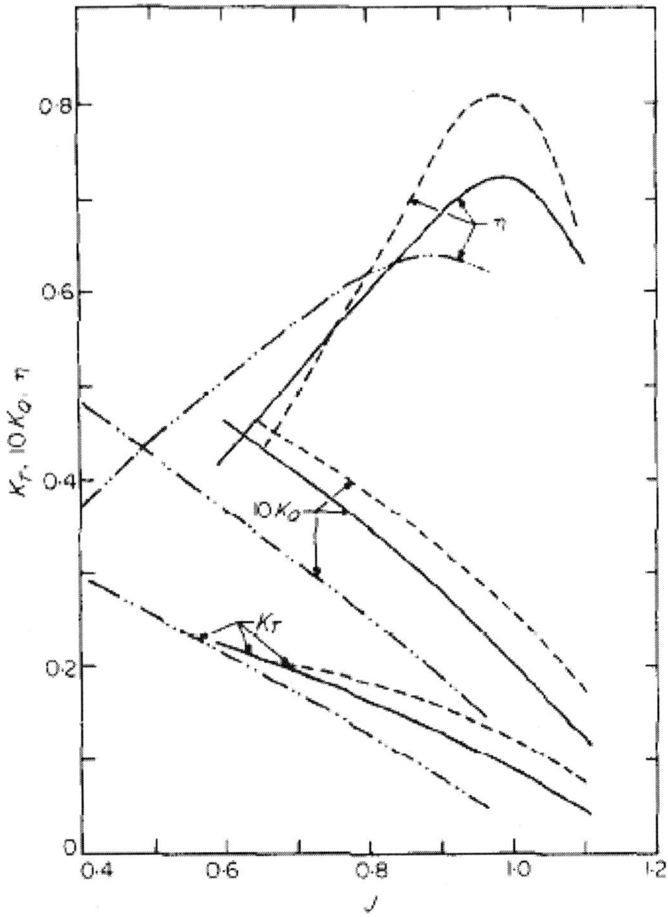

Figure 6. Performance diagram (Efficiency η, Thrust Coefficient K_T and Torque Coefficient K_Q for Propeller A (__), Propeller B (_ _) and Propeller C (_.._) vs. Advance Coefficient (J) $J = V / nD$

Based on the experimental results, the frequency range can be divided into two lower band frequencies (0 to 10kHz) and higher band frequencies (10 to 100kHz). According to the noise diagrams in two different operational conditions shown on Figures 5 and 6 and based on the hydrodynamic efficiencies, noise of the propellers can be explained.

As seen in Table 3, for the operational condition of K_T = 0.16, the efficiency of the Propeller C is lower than the other two types of propellers but from the hydroacoustic point of view, this propeller seems to be superior to Propellers A and B. It is clear that at the lower frequency range and at the frequency of 5 kHz, Propeller C generates 12dB less noise than the other two propellers. At the same time, it is shown that the increase of frequency will bring more reduction in the noise level of Propeller C. Propeller A and B have similar characteristics, but overall, Propeller B generates less noise. From the efficiency perspective and at the operational condition of K_T = 0.19, Propeller C stands in the middle and like before, it generates the lowest noise. A similar noise will be radiated from Propeller B at its highest efficiency level in high frequency range and Propeller A with its lowest efficiency will generate the most highest noise.

Table 3. Efficiencies of Propellers A, B and C in two different

KT	Efficiency		
	A(%)	B(%)	C(%)
0.16	59.7	72.5	57.6
0.19	51.6	57.2	53

In order to solve this problem, Sharma [12] modified the blades of Propellers A and B by drilling 300 holes with the diameters of 0.3mm. With this modification, a 50% reduction was achieved in propeller tip vortex which consequently resulted in a reduction in the propeller noise.

As evidenced in Figure 7, Sharma's modifications led to a decrease in cavitation and therefore a reduction of noise level.

Propeller's trailing edge (TE) geometry and its boundary layer play a significant role in the radiated noise of the propeller. Wang et al. [13] have done an extensive and precise study in 2006 on the effects of the propeller's trailing edge on the radiated noise. In this study, using numerical simulations, optimization of the trailing edge geometry was investigated. As a result of this optimization study, blades' trailing edge vortices and subsequently their generated noises were reduced especially in the case of low-frequency noise signatures. Results of this study are shown in Figure 8.

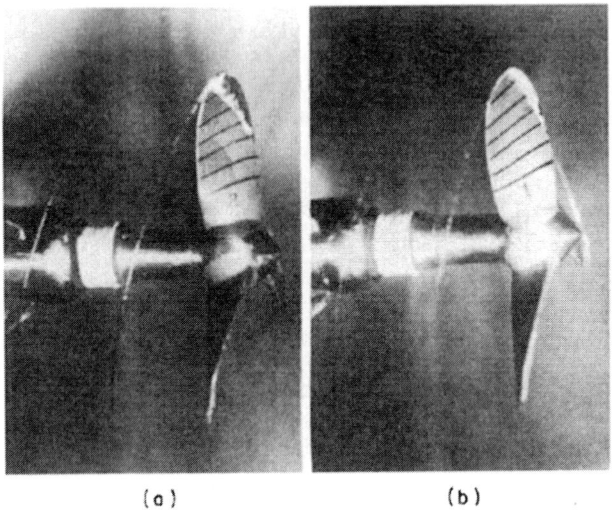

Figure 7. Propeller A in its original form (Left) and modified form (Right)

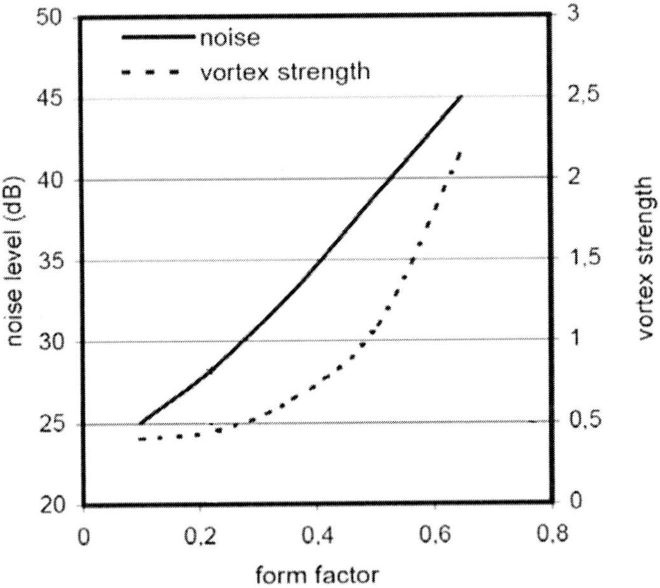

Figure 8. Trailing edge noise and vortex strength [circulation (m²/s] ($\Gamma = \int \omega da$) vs. trailing edge aspect ratio or form factor

It needs to be added that this study was actually a continuation of the Blake's [14] experiments in 1986 which was done on the effects of the local variations of the trailing edge aspect ratio in 2-dimensional configurations. In his studies, Blake varied the trailing edge from a very thick form to a very thin or sharp form. As observed in Figure 8, a sharper trailing edge will result in less noise radiation.

3. USING CONTRACTED AND LOADED TIP (CLT)

Another technique of noise reduction is the confinement of propeller's tip with a plane. An example of this method is shown in Figure 9. Confining the propeller's trailing edge will result in less circulation around the blade's tip and therefore the propeller will be put in heavier conditions. With this tool, diameter of the optimized propeller will be reduced and consequently the probability of cavitation occurrence will be decreased.

Figure 9. CLT propeller installed on a bulk carrier

Figure 10. Kappel Propeller installed on a ship

This system was first introduced by SISTEMAR Company and in an article by Gomez and Gonzalez [15] in 1995, the physical justification of this method was presented while the CLT theory was proven using the formulation and governing equations of the momentum theory.

In 2005, SISTEMAR performed experiments on two similar 164,000 tonnes bulk carrier ships and installed a new CLT propeller on one of them. Results showed that the ship with old type propeller had 12% more fuel consumption.

Also, the same system was installed on a passenger ship with twin screws propulsion system which used to suffer from vibrations problem [16]. This passenger ship had a length of 157 meters with 15,327 tonnes displacement. After the installation, interesting results were obtained. At speeds of 21 to 24 knots, 11% and at the speed of 15 knots, 30% reduction of engine power consumption was achieved. Additionally, propellers' pressure spectrum was reduced.

This type of propeller has many advantages but extra care should be given to its design.

4. KAPPEL PROPELLERS

Kappel propellers were first introduced by Andersen et al. [11] in 2000. In these types of propellers, the trailing edge is curved toward the suction region (back-face) and therefore in addition to reducing cavitation and respective cavitation noise, a 4% increase in the propeller's hydrodynamic efficiency can be obtained [16]. A typical Kappel propeller is shown on Figure 10.

5. PROPELLER HUB MODIFICATION

One of the propeller parts that play a significant role in introducing the pressure gradients and therefore the irregularities in the pressure distribution is the propeller's hub section. Effects of the blades are less in this section and therefore the velocity and pressure of water in this region is much lower than that of propeller's mid-regions. As a result, the hub section is exposed to cavitation and pressure oscillations. Consequently, this section should be given extra attention from cavitating and non-cavitating noise perspectives.

One of the methods that can be used in order to reduce the radiated noise in this section is the implementation of Propeller Boss Cap Fins (PBCF) on the propeller's hub as shown in Figure 11. With this tool, in addition to the decrease in fluid's circulation on the hub that will be achieved, hub cavitation can also be reduced [16].

This system has many advantages such as:

1. More efficiency (approx. 7%)
2. Cavitation tunnel test proved that sound pressure level (SPL) will be reduced by 3 to 6dB in frequencies higher than 1kHz
3. The ship does not need to be dry-docked for the installations of these fins and they can be fitted to propeller while the ship is still afloat
4. Use of PBCF will cancel hub vortex or hub cavitation (as shown on Figure 12) and therefore one of the major noise sources is attenuated

Figure 11. An example of using PBCF system on a marine propeller

Another cavitation and hub noise reduction method is the use of multiple hydrofoils on the hub which is called Propeller Cap Turbine (PCT). Similar to PBCF, PCT will cause less energy consumption while weighting around 1 to 2% of the total propeller's weight. A schematic of this system is shown [16] in Figure 13.

6. WAKE INFLOW MODIFICATION

Although the propeller noise is due to the propeller rotation and the consequent pressure distribution around it, propeller's wake inflow plays a significant role in the propeller performance and the probability of cavitation[17]. On the other hand, perturbation and fluctuations of free-surface can also affect the propeller's performance [18].

Figure 12. Ordinary propeller with hub cavitation (top) and the Propeller equipped with PBCF that cancels hub cavitation (bottom).

Therefore, in addition to the reduction methods mentioned in the previous sections, other methods are developed based on the modification of wake inflow. With the implementation of special elements that will be installed on the hull, a uniform inflow will be provided for the propeller which will abate propeller cavitation, propeller thrust oscillations and propeller vibrations. In addition to the increase in the propeller's efficiency, these methods will also decrease the radiated noise level.

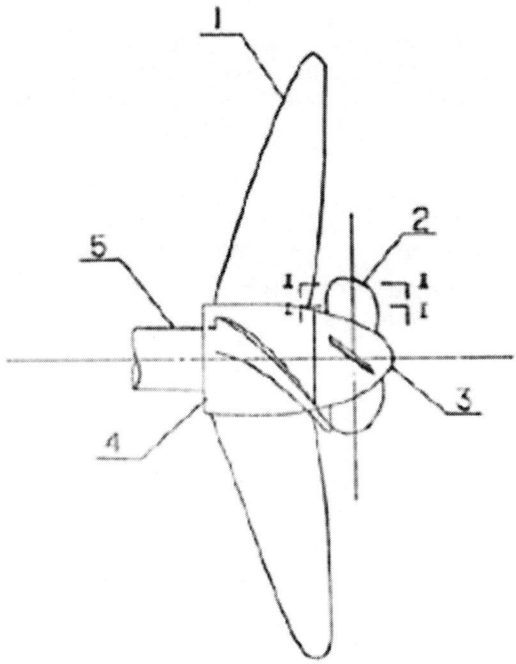

Figure 13. A schematic of PCT system

An example of these elements is the Schneekluth duct. The idea behind this element is developed in order to change the propeller's upstream flow which will result in less generated noise that will be radiated from the propeller's blade trailing edge. Consequently, the sound pressure level (SPL) will be decreased. However, there is not sufficient information about the effectiveness of this method in reducing the generated noise in all existing vessels. The Schneekluth duct will be installed on the hull of the ship and with the help of this element in addition to 12% decrease in fuel consumption, 50% less vibration will also be achieved. For example, the system has been installed on a 2500 tonnes Container ship with 22 knots speed. As a result of this modification, the annual fuel consumption has been decreased by up to 1200 tonnes. An example of this duct which is installed on a ship is shown [16] in Figure 14.

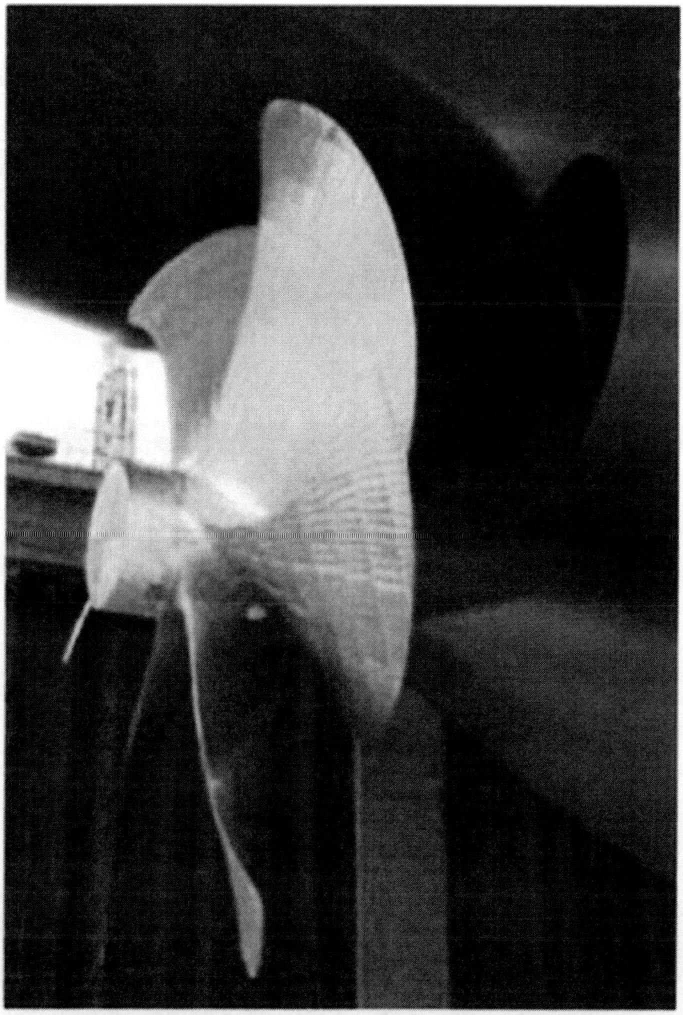

Figure 14. Schneekluth duct installed on a ship

Another element in this group of methods is the Mewis duct (shown in Figure 15). This element is placed just before the propeller and has been designed by Becker Marine Systems. Mewis duct will also bring better and more uniformed wake inflow for the propeller.

Figure 15. Mewis duct system

Figure 16. Simplified Compensative Nozzle or SCN system

Installation of this system on a VLCC ship with the length of 318m, breadth of 60m, water draft of 20m, 16 knots speed and 22,000 KW engine power has led to 5% less fuel consumption. Additionally, results show that with the modification of wake inflow, the efficiency of the propulsion system has been increased and propeller's cavitation and vibrations have been reduced [16].

Another method of modifying the wake inflow is called Simplified Compensative Nozzle or SCN. This tool is installed on both sides of the hull in the aft section as shown in Figure 16. The system in resemblance with the two latter methods will result in uniformity of the wake inflow in the stern region and will therefore reduce the generated noise due to the propeller's inflow oscillations. In addition to the advantages mentioned above, SCN system can be built a lot easier in compare to other inflow modification tools that were previously introduced[16].

7. PROPELLER'S CONFINEMENT: DIFFERENT TYPES OF DUCTS

Another family of methods that can be used in order to reduce the radiated noise from the propeller propulsion system both in marine and aviation applications is the use of ducts which surround the propeller. Generally, in aviation industries, two types of conical and cylindrical ducts are mainly used. With the implementation of these ducts, in addition to the propeller's efficiency increase, reduction in radial propagated noise will also be achieved. A practical way of improving acoustical behavior of these ducts is to use effective sound absorbers inside them.

In marine industries, ducted propellers are widely used. These ducted propellers are often called Kort Nozzle systems. Nowadays, different types of ducts are introduced with the most common types depicted in Figure 17. In general, there are two types of ducts in marine applications; accelerating and decelerating. The type that is shown in Figure 17-a is called Wageningen 19A which is among the most widely used ducts due to its high efficiency in forward thrust. This type of duct is suitable for vessels that do not require astern or backward sailing and cannot be effectively used for towing tugboats. For this reason, another type of duct (Figure 17-b) is introduced. For this type of duct, the astern sailing has more efficiency and on the other hand the efficiency for the forward sailing will be reduced. In the past two decades, investigations

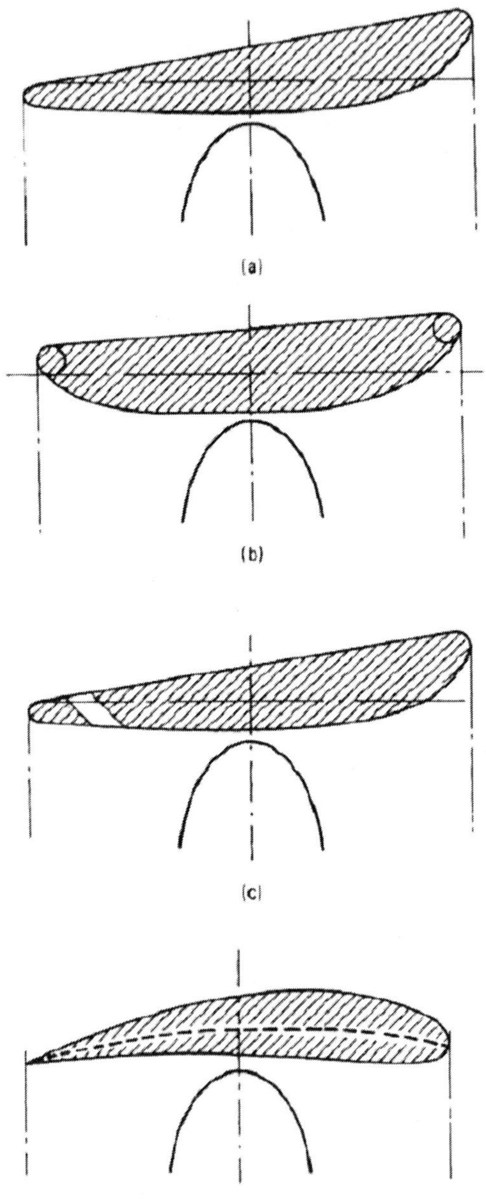

Figure 17. Different types of ducts used in marine industries for noise reduction applications

have been carried out for the improvements of efficiency in astern sailing for the ducted propellers. As a result of these investigations a new type of ducts (Figure 17-c) are introduced. In addition to the hydrodynamic section design in accordance with the forward sailing efficiency, these ducts take advantage of particular holes which are placed at the back of the ducts. In times of backward sailing, these holes let the water pass through and therefore the ducts will have the same performance as the first type of ducts [7].

For the sake of cavitation control and noise reduction, decelerating ducts are often used. This type of ducts which is illustrated in Figure 17-d improves the cavitation conditions by decreasing the propeller's inflow speed. These ducts are mainly used in naval ships, submarines and torpedoes.

In 2010, Haimov et al. [19] conducted some experiments on the performance and design methods of ducted propellers. Also, the studies of Seol et al. [20] in 2002 are among the most influential research works in this field. In their work [20], using a finite element method solver, they solved the potential flow around the propeller which later used as the input of the Ffowcs Williams-Hawkings (FWH) formulation. The FWH equations are solved using a boundary element solver and the far-field noise has been calculated for a typical propeller geometry. Later, using the presented method, the acoustical analysis has been performed on a Kappel Series propeller.

In the present section, different methods of reducing and controlling propeller noise were introduced. For a better overview of these methods and more practical applications, different classifications are presented in the following section.

8. CLASSIFICATION OF NOISE REDUCTION METHODS FOR MARINE PROPELLERS

For more efficient application of the noise reduction methods in the design of marine propellers, various classifications based on different perspectives are required. An example of these classifications was presented in the proceedings of the National Oceanic and Atmospheric Administration (NOOA) International Symposium in 2007 for large commercial ships [3], while this classification was mainly based on the implementation cost for each method.

Noise reduction methods can be classified in different categories. One of the suitable classifications is based on the type of the reduced noise, i.e.:

1. Cavitation Noise Reduction
2. Non-Cavitation Noise Reduction
3. Radiated Noise Reduction

Another important classification is based on the modification that is required for the implementation of the reduction method:

1. Propeller Wake Inflow Modification
2. Propeller Geometry Modification
3. Propeller Isolation

A combination of these two classifications are illustrated in Figure 18.

In addition to the classifications presented above, noise reduction methods can also be classified in terms of cost and required technology. If a score of 0 to 1 is assigned to each method in such a way that one (1) refers to the highest and zero (0) refers to the lowest level of cost/technology requirement, the diagram presented in Figure 19 will be achieved.

The vertical axis depicts the weighted parameters which account for each of the modification methods. The respective weights are estimated using the international figures for propulsion system installation/modification cost.

As evidenced in Figure 19, modifications in the propeller geometry is actually the most expensive method in sense of cost and technology with propeller isolation and wake inflow modification methods coming next. Propeller design is truly an intricate and expensive procedure and a slight fault in this phase can lead to less predicted efficiency in comparison to the standard calculations.

Noise reduction methods in sense of controlling particular frequency ranges are proved to be important and lead to a new classification of the methods. This classification includes low frequency and high frequency noise reduction methods which are depicted in Figure 20.

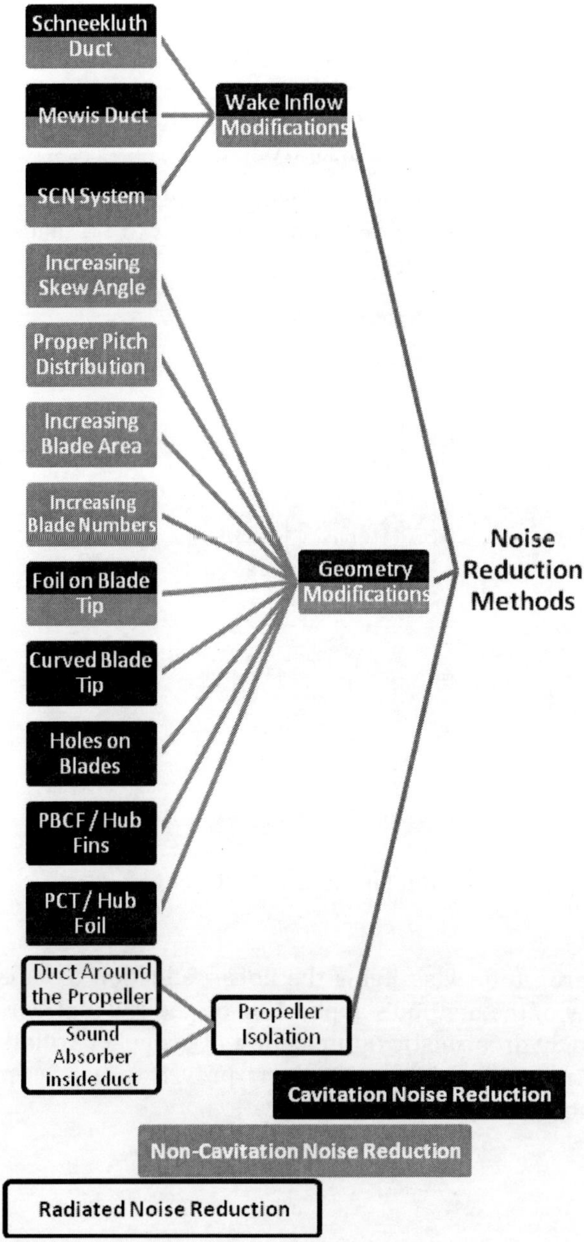

Figure 18. Noise reduction methods and their classification

High frequency noise reduction methods, generally attenuate or abate cavitation while the wake inflow or flow around the propeller are modified in low frequency noise reduction methods and the radiation of noise in this group of methods is controlled.

Based on these classifications, choosing a suitable reduction method would become easier when decisions are going to be made about the propulsion system of a vessel.

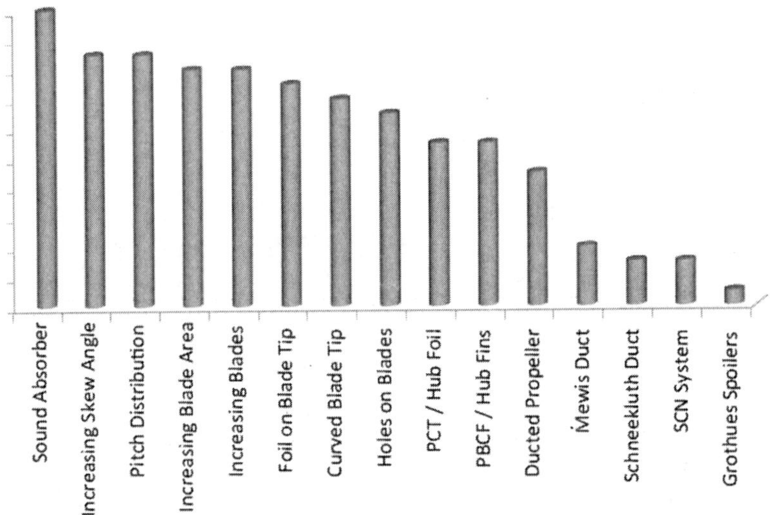

Figure 19. Noise Reduction Methods assortment

Another approach for classifying the noise reduction tools is based on the simplicity of the method's application on the existing vessels. In this classification, hydroacoustic optimization of propeller would be among the hardest methods while using ducts proves to be the easiest one. Figure 21 shows this system of classification.

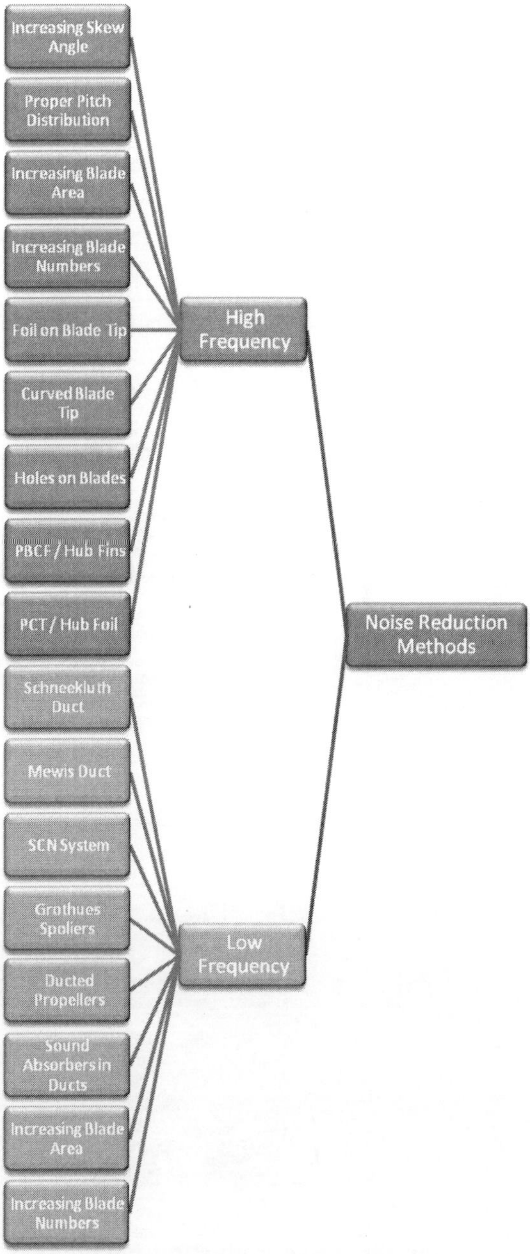

Figure 20. Classification of different noise reduction methods based on the frequency range of the attenuated noise

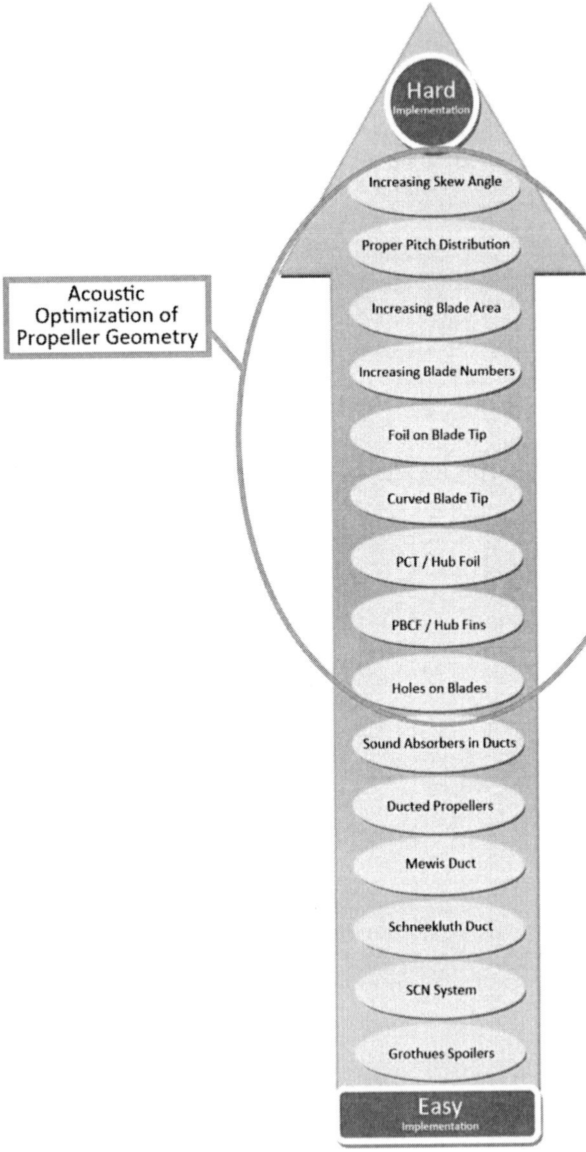

Figure 21. Classification of noise reduction methods based on the simplicity of application

9. CONCLUSIONS

In the present paper, principal methods of marine propeller noise reduction are presented and studied. These methods are then classified in five different categories which bring about better acquaintance with each of the methods, their limitations in cost and applications, which will help the designers make the right decision under different operational circumstances.

Studying different classifications proves that using ducts (in the aft region or around the propeller) is the easiest, most operative and most cost effective method in compare to the method of redesigning propellers for the existing vessels. Therefore, in order to control the radiated noise of the existing vessels, installation of ducts would be the first priority. This decision is due to the fact that by using ducts, radiated noise will be reduced, propeller's efficiency will be improved and the cavitation shedding will be delayed. Furthermore, using ducts can be the best and most effective choice for the vessels that are currently built and in operation. As the second priority, equalizing wake inflow with the help of aft ducts are presented which leads to reduced cavitation noise and improved propeller efficiency.

NOMENCLATURE

r radius

r_0 propeller's blade radius

C_m maximum camber length

c chord length

t_m maximum blade thickness

D propeller diameter

P propeller pitch

θ_{SP} propeller skew angle

K_T thrust coefficient

K_Q torque coefficient

η propeller efficiency

J advance coefficient

V ship speed

n propeller revolutions per second

dB decibel ($dB = 20\log\frac{P}{P_{ref}}$)

P sound pressure (Pa)

P_{ref} reference sound pressure level

(in water $P_{ref} = 1\mu Pa$, in air $P_{ref} = 20\mu Pa$)

Γ circulation (vortex strength) $\Gamma = \int \omega da$

SPL sound pressure level (dB)

REFERENCES

1. Ross, D., Ship sources of ambient noise, IEEE J. Oceanic. Eng. 30, 257-261, 2005.
2. Southall, B.L., Shipping Noise and Marine Mammals: a Forum for Science, Management, and Technology, Final Report of the National Oceanic and Atmospheric Administration (NOAA) International Symposium 2005, U.S.A. NOAA Fisheries, Arlington, Virginia, May 18-19, 2004, 40 pp.
3. Southall, B.L., Potential Application of Quieting Technology on Large Commercial Vessels, Final Report of the National Oceanic and Atmospheric Administration (NOAA) International Symposium. U.S. NOAA Fisheries, Silver Spring, Maryland, May 1-2, 2007.

4. Marine mammals and noise a sound approach to research and management, A report to Congress from the Marine Mammal Commission, USA, March 2007.

5. Shipping noise and marine mammals, A background paper produced by participants of the International Workshop on Shipping Noise and Marine Mammals, Okeanos Foundation for the Sea, Hamburg, Germany, 21st-24th April 2008.

6. Underwater Radiated Noise of Ocean-Going Merchant Ships, A background paper produced by participants of the International Workshop on Shipping Noise and Marine Mammals, Okeanos Foundation for the Sea, Hamburg, Germany, 21st-24th April 2008.

7. John Carlton, Marine propellers and propulsion, Elsevier, Second edition, 2007.

8. NVT Institute, Secrets of marine acoustical success cruising comfort., undated.

9. Matusiak, J., Pressure and noise induced by a cavitating marine screw propeller, Doctor of Philosophy thesis, VTT Publications 87, 1992.

10. Koushan, K, Halstensen, SO, and Sandtorv, LG, 2000, Systematic investigation of blade design influence on cavitation performance and on induced pressure pulses, Proceedings of NCT 50, International Conference on Propeller Cavitation, 3-5 April 2000, University of Newcastle, UK.

11. Andersen, P., Kappel, J.J., Spangenberg, E., Aspects of propeller developments for a submarine, First International Symposium on Marine Propulsors, smp'09, Norway, June 2009.

12. Sharma, S. D., Mani, K., Arakeri, V. H., Cavitation Noise Studies on Marine Propellers, Journal of Sound and Vibration Vol. 138, 255-283, 1990.

13. Wang, M., Marsden, A.L., Moin, P., Computation and Control of Trailing-Edge Noise, Proceedings of the 26th Symposium on Naval Hydrodynamics, Rome, 2006, Italy.

14. Blake, W. K., Mechanics of flow-Induced Sound and Vibration, Vol. II, Complex Flow-Structure Interactions, Applied Mathematics and Mechanics, 1986, Academic Press.

15. Gomez, G.P., Gonzalez-Adalid, J., Tip loaded propellers (CLT): Justification of their advantages over conventional propellers using momentum theory, International Shipbuilding Progress, 42 no 429, pp 5-60, 1995.

16. Renilson Marine Consulting Pty Ltd., Reducing underwater noise pollution from large commercial vessels, Commissioned by The International Fund for Animal Welfare, March 2009.

17. Ball, W., An experimental investigation into the influence of wake on cavitation, Transactions of the Royal Institution of Naval Architects, Volume 131, pp 73-82, 1989.

18. Arndt, REA, Holl, JW, Bohn, JC, Bechtel, T, Influence of surface irregularities on cavitation performance, Journal of Ship Research, Volume 23, Number 3, September 1979, pp 157-170, 1979.

19. Haimov, H., Bobo, M. J., Vicario, J., del Corral, J., Ducted propellers: a solution for better propulsion of ships. calculations and practice, First International Symposium on Fishing Vessel Energy Efficiency, E-Fishing, Vigo, Spain, May 2010.

20. Seol , H., Jung , B., Suh , J.-C., Lee S., Prediction of non-cavitating underwater propeller noise, Journal of Sound and Vibration, Vol. 257, 131-156, 2002.

6

CHAPTER

Effects of Marine Propeller Performance and Parameters Using CFD Method

Kiam Beng Yeo, Rosalam Sabatly, Wen Yee Hau and Cheah Meng Ong

Materials and Minerals Unit, Faculty of Engineering, Universiti Malaysia Sabah, Jalan UMS, Kota Kinabalu, 88400, Sabah, Malaysia

ABSTRACT

This study presents the investigation of marine propeller hydrodynamic performance and parameters through Computational Fluid Dynamic analysis. Propellers with different Pitch to Diameter (P/D) ratio, propeller blade number (Z), skew angle, rake angle and Expanded Area Ratio (EAR) were subjected to computational flow analysis based on the Reynolds Averages Navier-Strokes Equation-(RANSE) solver. Results found that all thrust coefficient (KT) and torque coefficient 10 (KQ) decreases with the increasing advance coefficient (J). The efficiency of propeller performance had also consistently showed characteristic trend of nonlinear increases to a peak an optimum value before decreasing drastically with increasing J value. The analysis found that no single marine propeller achieve optimum performance in thrust, torque, efficiency and velocity with less loading on the mechanical properties.

INTRODUCTION

A good propeller is required to produce the best hole-shot, high stability and efficiency for a ship movement to be adequate in the range of operations either fast speed and light load, or low speed and heavy load.

Nevertheless, manufacturer's propeller series are vast and complex to be fully understood without real involvement of users in the development. Researches are continuously working on the entire reasons and effects of marine propeller to solve for the enhancement of hydrodynamics flow behaviors by refining the propeller shape profile.

Mihaela (2005) and Miyata (1997) found that Computational Fluid Dynamics (CFD) of Solidworks had a numerical algorithm capable of performing hydrodynamic analysis. It decomposed the spatial domain into tiny cells to form the volume mesh and a suitable algorithm was applied to solve the equation of motion (Wu, 2010). In Bertram (2010) research, however, he sectioned the flow analysis of CFD into two parts, which are computer based simulation of fluid flow modeled by field equations and fluid domain. On the other hand, Subhas *et al.* (2012) computed an estimation of thrust and torque performance for different rotational speeds found an extremely ideal CFD result when compared with the experimental work.

Liu *et al.* (2012) had also suggested that uncertainty in CFD analysis had lowered the thrust and torque performance by 5% could be utilized as the standard deviation of measurement of computational value. They further suggested that BEM code has the most reliable method in obtaining potential flow results. They attributed the variance in CFD result as due to software code lacking grid density or too much of numerical dissipation in the vicinity of cavity-fluid interface.

PROPELLER PERFORMANCE ANALYSIS, PARAMETERS AND SETTINGS

Marine propeller hydrodynamic characteristics: Hydrodynamic flow characteristics are important to the complete design of propellers as well as the ship. The hydrodynamic interactions with the propeller provided a series of phenomenon that are related to the **physical properties**. Dimensionless analysis of propeller performance in terms of thrust coefficient (K_T), torque coefficient (K_Q) and efficiency (η) with respect to advance coefficient, J are standard parameters to evaluate the efficiency of a propulsion system (Mehdi *et al.*, 2010) related as:

$$K_T = \frac{T}{\rho N^2 D^4} \tag{1}$$

$$K_Q = \frac{Q}{pN^2D} \quad (2)$$

$$J = \frac{V_a}{ND} \quad (3)$$

where, T is the Thrust, Q is the Torque, P is the Density of fluid, N is the Rotational speed, D is the Diameter and V_a is the Speed of advance. According to Carlton (2007), propeller performance efficiency could be normalized as an optimum curve indicating increasing trend and decreases over a peak value. In most studies, J as the forward movement of propeller with respect to its diameter and rotational speed to determine the efficiency, thrust and torque coefficient was used. The performance was normalized as a characteristic efficiency to identify the optimum parameters for the propeller. Further, he also showed that the values of K_T and K_Q were decreased with with increasing J. The result of John showed similar findings by other researchers on other marine propeller series and experimental models (Mehdi *et al.*, 2010).

Computational analysis settings: The CFD analysis and simulation of marine propeller performance under hydrodynamic fluid interaction was conducted. The propeller geometry was generated according to specification before CFD flow simulation set up. An accurate and complete path of setup was required to obtain relevant simulation result. Selection excluding cavities and internal space also reduces time consumption in the CFD computation. Since the actual operating environment of marine propeller was seawater, hence, dynamic similarities were selected at 24°C with a viscosity of 9.11×10^{-4} pa.sec. In the analysis, wall roughness was assumed at 0 μm or smooth, which excluded the pollutant dispersion and deposition; assuming ideal flow characteristic. Thermodynamic parameter of initial standard atmospheric pressure of 101,325 Pa with activation of pressure potential allows incremental computation of depths in the domain of marine propeller. Also, the temperatures of fluid and solid were adjusted according to the actual investigation condition.

Solidworks Flow Simulation does not provide initial setup of solid Computer Aided Design (CAD) model with velocity parameter, except for rotating regions and velocity of flow, which acts perpendicular to the back surface of marine propeller controlled by domain wall initial speed. This was sufficient for propeller-fluid interaction to assimilate the real operation condition. Result and geometry resolution by initial setting allows control of analysis and meshing accuracy, which was time and memory capacity dependent. The governing solution accuracy grossly depends on the geometry resolution governed by computational mesh settings in the minimum gap size, minimum wall thickness and mesh number in the computational domain.

PROPELLER FLOW VELOCITY PERFORMANCE AT DIFFERENT RPM

Flow velocity of different Z at various RPM: Flow velocity against rotational speed with number of blades at Z = 2, 3, 4 and 5 of Fig. 1, Z = 5 reached 18.4 m sec^{-1} and Z = 2 at 15.7 m sec^{-1} in 6000 rpm consisting of 17.2% of variance displacement per unit of time. At initial startup, the higher the Z number, the more rotation was required to achieve the equivalent speed. However, after attaining steady rotation, higher Z emerged as the best blade numbers that bites most water in providing the highest velocity for each rpm. Although higher Z provided better velocity, but it leads to higher fuel consumption and so reduce the efficiency (η).

Flow velocity of different EAR at various RPM: For the velocity and RPM with different EAR (Fig. 2), the smallest value being related to the lowest flow velocity of EAR = 0.3 and 15.7 m sec^{-1} followed by EAR = 0.4 at 16.3 m sec^{-1}, EAR = 0.5 at 17 m sec^{-1}, EAR = 0.6 at 17.6 m sec^{-1} and highest with EAR = 0.7 at 17.9 m sec^{-1}. Increasing the EAR from 30-70% could only increase 14% of flow velocity, while increasing the propeller swipe area by 133%. In terms of geometry, higher the EAR means a heavier weight and more fuel consumption for each rotation.

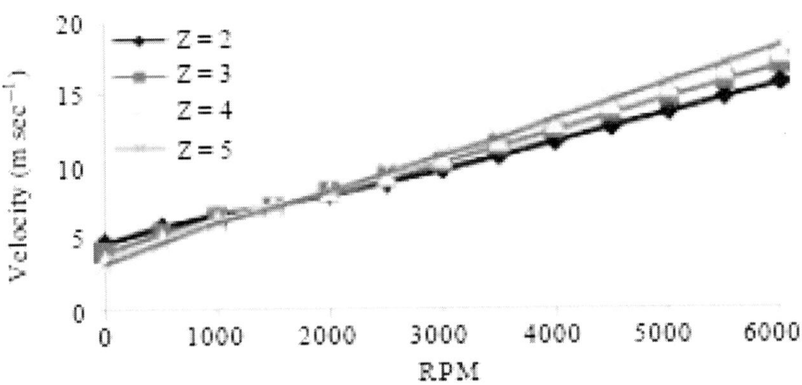

Figure 1: Flow velocity against RPM for different Z

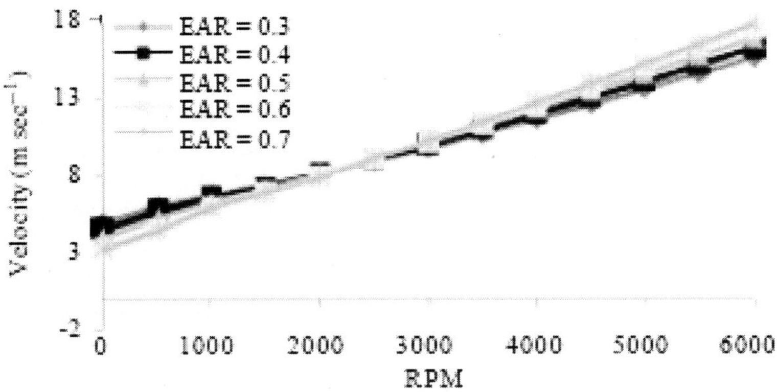

Fig. 2: Flow velocity against RPM for different EAR

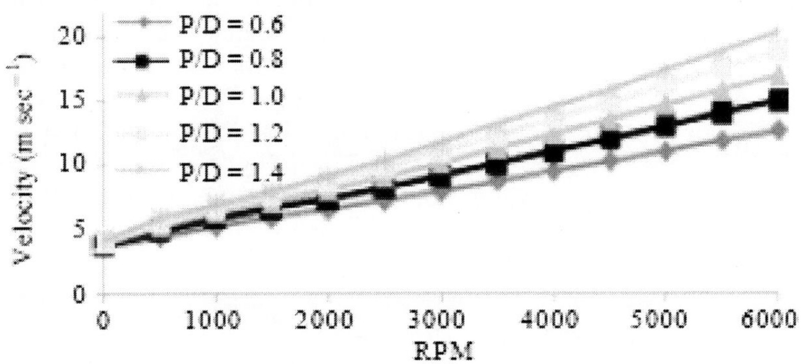

Fig. 3: Flow velocity against RPM for different P/D ratio

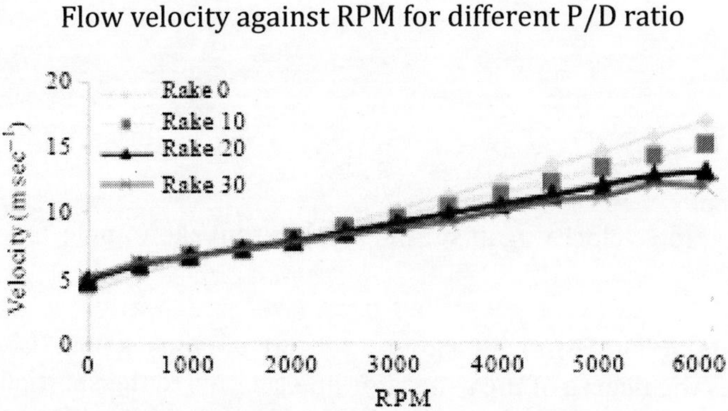

Fig. 4: Fow velocity against RPM with different rake angles

Flow velocity of different P/D ratio of propellers at various RPM: In Fig. 3, linear increasing pitch-diameter ratio occurs with increase rotational speed. Fluid to solid blade surface interaction at zero rpm had a low P/D = 0.6 at 3.6 m sec^{-1} and highest P/D = 1.4 at 4.2 m sec^{-1}. Increasing P/D ratio with higher startup velocity was attributed to the inclined pitch angle to horizontal axis, where blades mounted on the root had reduced the projected area to cause lower thrust of fluid flow acting on the blade. The optimum range of operation speed in most commercial propeller, P/D = 1.4 had achieved the highest speed of 20.33 m sec^{-1} and P/D = 0.6 the lowest at 12.7 m sec^{-1}.

Flow velocity of different rake angle at various RPM: Rake angle measures the degree propeller blade-angle being perpendicular to the hub. Blade rake increases as the blade-face slants backward to the propeller rear ship aft, which was positive rake angle were studied. Results, showed lower rake angle of 0° and 10° linearly increased with higher rpm (Fig. 4). As rake angle increased at higher rotational speed, the numerical values of velocity became unstable. Previous studies suggested that tunnel boats and air entrapment hulls could be unstable for high rake propeller and most propellers had rake angle of 0-20° only.

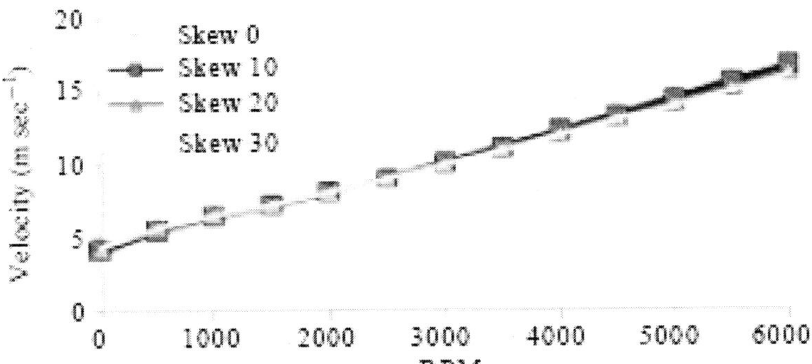

Fig. 5: Flow velocity against RPM for different skew angle

Flow velocity of different skew angle at various RPM: Skew angle measures the degree of the generator line relative to the shaft of a blade section. Velocity of marine propeller with respect to different skew angle of 0-30° had shown an increasing velocity curve with respect to the RPM. Net decreasing velocity with increasing skew angle in propeller design had not provided higher velocity value in this analysis. Instead, the changes of skew had only caused small velocity reduction over the whole operation range (Fig. 5).

PROPELLER PERFORMANCE AND OPERATING PARAMETERS

For convenient comparison between one propeller geometry to another, propeller performance had been defined in non-dimensional parameters. From the particular operating parameters with respect to one another, it was recommended to combine all these parameters in a master graph presented in the following results.

$10K_Q$, K_T and η for various Z with different J: From the CFD analysis, changes of blade numbers for standard propeller geometry of EAR = 0.5, P/D = 1.0, 0° rake angle and 0° skew angle showed that all propeller optimum η values occurred at about 0.78 of J (Fig. 6). For Z = 2, peak performance was achieved at more than 63%, Z = 3 at 61.3%, Z = 4 at 60.8% and Z = 5 at 57.6%. For $10K_Q$ at peak η, Z = 2 was achieved at more than 19.4%, Z = 3 at 24.5%, Z = 4 at 29% and Z = 5 at 30.85%. On the other hand, K_T at peak η for Z = 2 achieved at 9.4%, Z = 3 at 11.9%, Z = 4 at 13.5% and Z = 5 at 13.95%. For the best Q to η and T to η values, it was only relevant at the point when decreasing curve $10K_Q$ and K_T first hit the η curve. The Q, T and η best intersection point for each Z-parameter were shown in Table 1. All Z had η curves showed rapid increase from 0 rpm and decreased steadily after the peak η, while 10 K_Q and $K_T$decreases as J starts to increase.

$10K_Q$, K_T and η for various EAR with different J: Marine propeller profile for Z = 3, P/D = 1.0, 0° of skew angle and 0° of rake angle with changes of EAR from 0.3-0.7 were conducted, Fig. 7. The performance curve of best η for EAR = 0.4 had the highest 62.4% followed by EAR = 0.5 at 61.25%, EAR = 0.3 at 59.22%, EAR = 0.7 at 58.37% and EAR = 0.6 at lowest value of 56%. Best performance for intersection of $10K_Q$, K_T, η was also displayed in Table 2. All trend of η curves were increasing to peak at J = 0.8 before decreasing drastically.

Table 1: Best performance of Q-T-η parameters for different Z

Z	Q-η	T-η
2	38.2% at J = 0.379	27.2% at J = 0.259
3	43.6% at J = 0.452	30.9% at J = 0.313
4	46.9% at J = 0.498	33.4% at J = 0.339
5	48.5% at J = 0.529	34.6% at J = 0.359

Table 2: Best performance of Q-T-η parameters for different EAR

EAR	Q-η	T-η
0.3	39.6% at J = 0.398	27.8% at J = 0.272
0.4	42.1% at J = 0.427	29.5% at J = 0.293
0.5	43.6% at J = 0.452	30.8% at J = 0.313
0.6	44.7% at J = 0.486	31.8% at J = 0.331
0.7	47.1% at J = 0.473	34.3% at J = 0.342

Also, Q and T performances were decreased with the increment of J with higher EAR decreased much faster than lower one. $10K_Q$ and $K_T$values were initially higher at larger EAR ratio, but smaller EAR profiles took over larger ones after crossing the peak efficiency curve at J = 0.8.

$10K_Q$, K_T and η for various P/D ratio with different J: From the Fig. 8, each P/D ratio profile had a peak η with respect to the J. Instead, $10K_Q$, K_T value for all P/D profiles were varied with different values over the η curve and decreased smoothly with respect to J. As the P/D ratio went higher, the efficiency value also increased; highest efficiency for various P/D ratio in this result was 1.4 achieved at more than 67%, followed by P/D = 1.2 at 61.5%, P/D = 1.0 at 61%, P/D = 0.8 at 59% and the lowest was P/D = 0.6 at 54%. The trend of $10K_Q$ and K_T to η, low P/D ratio propeller intersected the η curves at lower figures and higher P/D ratio to achieve at greater one. The optimum performance of Q and T when it first intersected the η curve was displayed in Table 3.

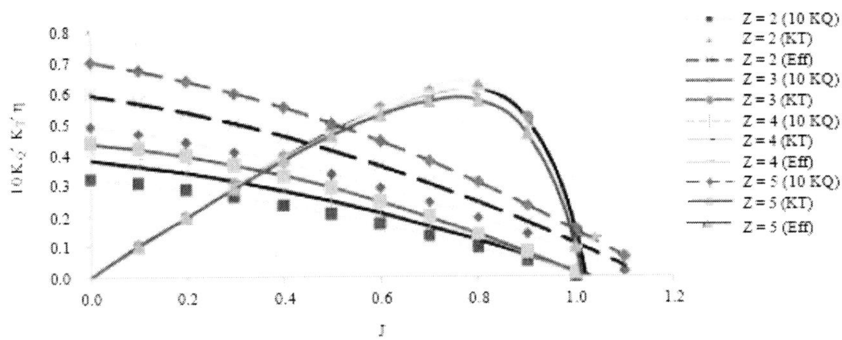

Fig. 6: Graph of 10 K_Q, K_T, η with respect to J for different Z

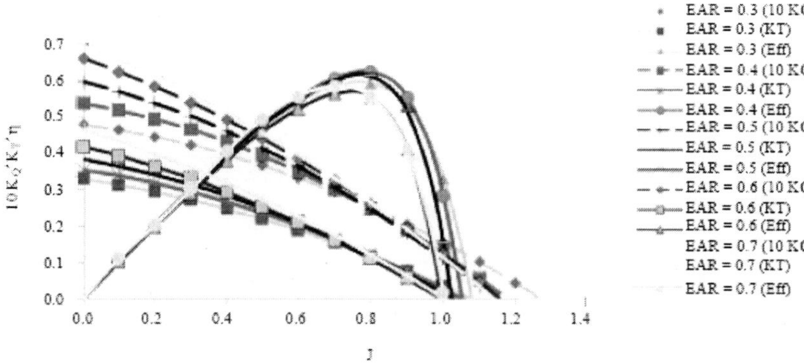

Fig. 7: Graph of 10 K_Q, K_T, η with respect to J for different EAR values

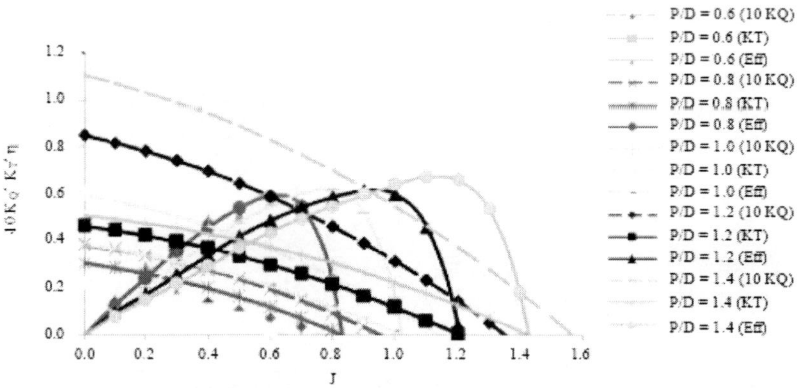

Fig. 8: Graph of 10 K_Q, K_T, η with respect to J for different P/D ratio

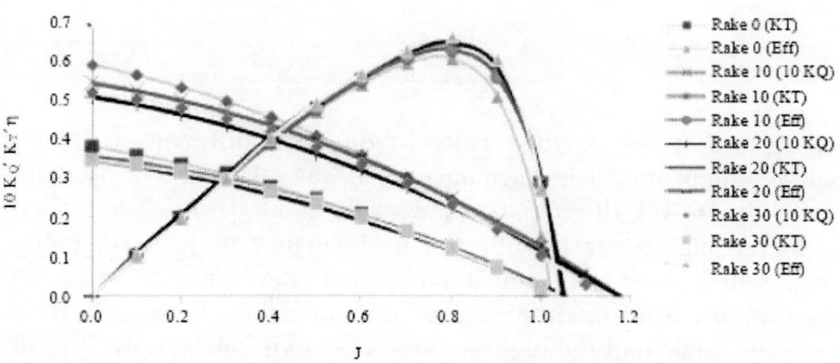

Fig. 9: Graph of 10 K_Q, K_T, η with respect to J for different rake angle

Table 3: Best performance of Q-T-η parameters for different P/D ratio

P/D	Q-η	T-η
0.6	20.72% at J = 0.127	19.11% at J = 0.131
0.8	31.81% at J = 0.266	25.57% at J = 0.213
1.0	43.63% at J = 0.452	30.91% at J = 0.312
1.2	53.34% at J = 0.688	35.82% at J = 0.425
1.4	60.38% at J = 0.924	39.17% at J = 0.537

Table 4: Best performance of Q-T-η parameters for different rake angle

Rake (°)	Q-η	T-η
0	43.63% at J = 0.452	30.88% at J = 0.313
10	42.57% at J = 0.432	29.99% at J = 0.298
20	40.5% at J = 0.3950	29.14% at J = 0.290
30	41.81% at J = 0.412	29.61% at J = 0.296

Table 5: Best performance of Q-T-η parameters for different skew angle

Skew Angle (°)	Q-η	T-η
0	43.63% at J = 0.452	30.9% at J = 0.313
10	44.11% at J = 0.457	31.4% at J = 0.316
20	44.65% at J = 0.465	31.5% at J = 0.318
30	45.55% at J = 0.467	32.0% at J = 0.324

10K_Q, K_T and η for various rake angle with different J: The Fig. 9 analyses the propeller performance for 0-30° rake angle of propeller profile with Z = 3, EAR = 0.5, 0° skew angle and P/D=1.0. The η curve were increasing to reach optimum performance at J = 0.8, before drastically decreased to J value at 1.1. Higher skew angle had achieved the best η but not for the intersection point for 10 K_Q, $K_T$and η, Table 4. The 0° rake angle had the highest value of Q to η followed by 10°, 30° and finally 20° with the lowest net Q value. The phenomena of Q to η also similarly occurred to the K_T characteristics curve. Table 4 displayed the best Q and T results at the first intersection with the η curve.

$10K_Q$, K_T and η for various of skew angle with different J: The Fig. 10 with various skew angles for Z = 3, EAR = 0.5, P/D = 1.0 and 0° of rake angle had increasing efficiency before drastic decreased right after the J = 0.8, while the 10 K_Q and K_T decreased with increasing J values. At 30° of skew angle, the highest η value was achieved at 65%, followed by skew 20° with 62.97%, skew 10° with 62.55% and the lowest was zero skew angle with 61.25%. Best performance for intersection of 10 K_Q, K_T and η was displayed in Table 5.

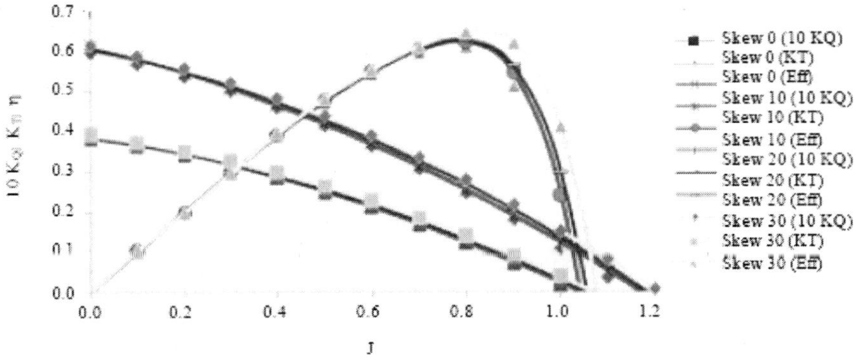

Fig. 10: Graph of 10 K_Q, K_T, η with respect to J for different skew angle

CONCLUSION

The performances and parameters investigation were conducted by implementing the computational fluid dynamics and **finite element**analysis to a standard set of marine propeller geometry. Results had shown that all K_T and 10 K_Q dropped with the increasing of J values. Furthermore, the efficiency of the propeller had increased at a small inclination before an optimum peak value was achieved and then to drop drastically right after with increasing Advance Coefficient J. Finally, no single marine propeller could achieve a single optimum performance in thrust, torque, efficiency and velocity with less loading on its **mechanical properties**.

ACKNOWLEDGEMENT

The authors would like to express their appreciation to the Materials and Mineral Unit, Faculty of Engineering, Universiti Malaysia Sabah and the Ministry of Higher Education of Malaysia for the financial support through the research grant FRG0249-TK-2/2010 and FRG0247-TK-2/2010.

REFERENCES

1. Bertram, V., 2010. Appropriate tools for flow analyses for fast ships. Germanischer Lloyd, Hamburg, Germany.
2. Carlton, J.S., 2007. Marine Propellers and Propulsion. 2nd Edn., Butterworth-Heinemann, Oxford, UK.
3. Liu, D., F. Hong, F. Zhao and Z. Zhang, 2012. The CFD analysis of propeller sheet cavitation. China Ship Scientific Research Center, Wuxi, China.
4. Mehdi, N., M.J. Abbasi and A.M. Amini, 2010. Assessment of marine propeller hydrodynamic performance in open water via CFD. Proceedings of the International Conference on Marine Technology, December 11-12, 2010, Dhaka, Bangladesh, pp: 35-44.
5. Mihaela, A., 2005. Developments in the design of ship propeller. Scientific Bulletin of the Politehnica University of Timisoara Transactions on Mechanics, University of Galati.
6. Miyata, H., 1997. Time-marching CFD simulation for moving boundary problems. Proceedings of the 21st Symposium on Naval Hydrodynamics, June 24-28, 1996, Trondheim, Norway -.
7. Subhas, S., V.F. Saji, S. Ramakrishna and H.N. Das, 2012. CFD analysis of propeller flow and cavitation. Int. J. Comput. Applic., 55: 26-33.
8. Wu, X., 2010. A Rapid Development Process for Marine Propellers through Design, Simulation and Prototyping. Memorial University of Newfoundland, St. John's, Newfoundland and Labrador, Pages: 244.

7
CHAPTER

Determination of Optimal Stacking Sequence for Modal Characteristics Evaluation of Composite Marine Propeller Blade

M.L. PavanKishore[1], R.K. Behera[1]

[1]*Department of Mechanical engineering, National Institute of Technology, Rourkela, India*

ABSTRACT

The design of optimum marine propeller is one of the most important aspects of naval architecture. With the increase in demands for high operating efficiency, power and low level of noise, vibration reduction the design of propellers became extremely complex. This paper describes the numerical prediction of free vibration characteristics of a B-series propeller using finite element approach as a base line method. The propeller analysis is performed as a single objective function subjected to the constraints imposed by cavitation, material strength and propeller thrust. An important aspect of autonomous underwater vehicle is to evaluate its modal characteristics in terms of its mode shapes and natural frequencies. The effect of stacking sequences, fibre orientation angles are studied and finally an optimum stacking sequence has been determined for optimum characteristics of B-series (B4-0.7) marine propellers.

Keywords: Ansys, finite elements, mesh, mode shapes, pre twist

1. INTRODUCTION

Since the advent of propellers blades in the application of marine engineering cause an important role. The blade failure is predominantly vibration related it may be either marine propeller blade or steam turbine blade or wind turbine blade. The blades being very flexible structural members in the sense that a significant number of their natural frequencies can be in the region of possible nozzle excitation frequency. From manufacturers point of view it is important to know the natural frequencies of the blades, in order to make sure that the boss on which the blade should not have some of the frequencies otherwise a resonance may occur in the whole structure leading to un damped vibrations which may eventually wreck the ship.

The movement of a ship through the water is achieved by the power so developed in the engine via the propeller shaft to propel in water. The propeller produce thrust through the production of lift by their rotating blades. The blades are attached to boss or hub. The boss is fitted to the propeller shaft through which the power of the propulsion machinery of the ship is transmitted to the propeller. The strength requirements of propellers indicate that not only the blades be sufficiently strong to withstand long periods of arduous service without suffering failure. J.E.Collony [1] address the problem of the wide blades, tried to combine both theoretical and experimental investigations. A single free standing blade can be considered as cantilever beam with a rectangular cross section. The vibration characteristics of such blade are always coupled between the two bending modes in the flap wise and chord wise directions and the torsion mode. Leissa and Jacob [2] were the first ones to investigate the free vibration characteristics of cantilevered twisted beams and plates as a three dimensional vibrational problem. Kielb [3] presented a comprehensive review of structural dynamics aspects of pre twisted beams following on the same path of lines presented very comprehensive summary of the analytical work performed in the area of twisted plates. The equation relating to twisting plates are much more complex than corresponding Euler and Timoshenko beam formulations. An early analytical model to calculate the natural frequencies of a cantilever beam was suggested by South well [4] based on Rayleigh energy theorem which defines a simple equation relating the fundamental frequency to that of rotating frequency of a beam was suggested. Fried man & Straub [5] have formulated the discretization of linear equations of motion of blade by the Galerkin method and considered the problem of aero elasticity based on the finite element method of variable order. Sivaneri& Chopra [6], Thakkar & Ganguli [7] treated Hamilton's principle as a variation principle for the rotor blades analysis. Crespo da silva [8] established a partial derivative nonlinear rotor movement and vibration of the blade taking into account of the geometric non linearity and analyzed the dynamic stability of the vibration movement. Comparing the beam model, plate and shell are more approximate models of blades. Ramamurti and Kielb [9] studied rotating twisted plates by using finite element method with two different shape functions. Sreenivasa murthi and Ramamurti [10] studied rotating pre twisted and tapered plates by finite element method with triangular shell elements having 3 nodes with 18 degrees of freedom respectively. Leissa et al [11] studied cylindrical shells with chamber and twist on the shallow shell theory by Ritz method. Recently Tsuji et al [12] studied rotating thin twisted plates by the Rayleigh Ritz method.Hu and Tsuji [13] studied the rotating thin cylindrical panels with twist on general shell theory. Lo [14] in his analysis simplified the nonlinear problem by assuming the blade to be rigid everywhere except at the root presented the solution in phase plane. Boyce, Diprima& Handle man [15] used Rayleigh Ritz and South well methods to determine the upper and lower bounds of natural frequencies of turbine blade vibrating perpendicular to plane of rotation. Schihans [16] investigated the stiffening effect of the centrifugal forces on the first mode frequency of a rotating cantilever blade by the use of successive approximation. Hirsch [17] considered the effect of non-rigid supports. Horway [18] considered hinged ends and Nodorson [19]considered the effect of loose hinged supports on the natural frequencies of rotating blades. Garland [20]considered the case of a cantilever blade and used the Rayleigh Ritz method to determine the frequencies and amplitudes.

1.1. Composite

The implementation of fiber reinforced composite materials to the application of blades continues to increase, since it is important to explore the potential benefits that can be designed into the physics of these materials. Vibration is often critical to the successful operation of engineering structures which are composed of composite materials for example propeller blades, helicopter rotor, wind turbine blades, automatic and aerospace panels. Most of these structural components can be approximated as laminated composite beams[21].Free vibration analysis of laminated composite beams has been conducted by significant amount of research. Yddorom & Koral [22] studied the out of plane free vibration problems of symmetric ply laminated beams using the transformation cross method Khedir & Reddy [23] have been studied free vibrations of cross ply laminated beams with arbitrary boundary conditions. Investigations on the natural frequency modes of graphite epoxy cantilever plate and shells was carried out by crawley [24] and the free vibrations of rotating composite plates was analyzed by Wang [25] and Shaw [26].The first established work on pre twisted composite plates was carried out Qatu et al [27] to determine the natural frequencies of stationary plates using laminated shallow theory using Ritz method. Brig and Migliore [28] provided preliminary design method based on theory of Euler Bernoulli theory on layup structures and it was found that layup sequence which was obtained by this method did not meet the requirements of actual strength at the root area through analyzing the finite element model. Alejandro [29] proposed to analyze basic vibration mode of composite layup structure of wind turbine blades with the normal operating conditions by transforming complex geometric blade model into equivalent beam model using the VABS calculating program Wangyu Liu [30] studied the influence of fiber angle of the layup blade on strength through the finite element model of the blade and combined the response surface methodology and results showed that when layup angle is near 45 it can get higher stability strength. Stacking sequence of a laminated cylindrical shell is optimized based on natural frequency by Shakeri et al[31].

2. MATERIALS OF PROPELLER BLADE

The materials from which the propellers are made in present can be broadly classified as members of bronze or stainless steels. These materials are better known materials that are in use for manufacture of all types of propellers ranging from the large commercial vessels and warships through to pleasure run about and modeled propellers. The use of high tensile Brass accounted for 64% of all propellers produced with Manganese Aluminum Bronze and Nickel Aluminum Bronze accounting for small proportions in 1960's. In 1980's NAB has completely gained dominance over the other materials accounting for 82% of the propellers. In present trend high tensile brass accounts for less than 7% of the propellers used. The stainless steel gained comparatively popular usage in the period from 1960 to 1970's but progressively lost favour to the copper based materials to account for

3% of the materials used for propeller manufacture and are mostly applied to ice class propellers. properties of both materials are given in the following tables.

2.1. Carbon Based Composites

In recent years carbon based composites as propeller material have made an entry into the commercial craft and yacht market. Since they have been used in specialized naval vehicles such as submarines for many years. In addition to the use of acoustic properties for the materials there is a further advantage in their light weight when compared with conventional marine propellers.

Table 1. Material properties for metallic propeller

Material used	MAB
Young's modulus	125.5[Gpa]
Rigidity Modulus	44[Gpa]
Poisson's ratio	0.326
Density	7.53[g/cc]

3. SOLID MODLEING OF PROPELLER BLADE

Propellers more particularly propeller blades are more complex shapes which require the right hydrodynamic surfaces for modeling. The important parameters that characterize the modeling of a propeller are thus diameter, number of blades, the type of series for blade section, pitch distribution, the skew angle, the rake angle, the area of the blades. In this work a four bladed propeller was chosen for analysis. The global parameters that define the modeling of propeller blade are given as follows

Table 2. Material properties of composite propeller

Material	S-glass fabric/epoxy
Ex	22.925[Gpa]
Ey	22.925[Gpa]
Ez	12.4[Gpa]
υ_{XY}	0.12
υ_{YZ}	0.2
υ_{ZX}	0.2
Gxy	4.7[Gpa]
Gyz	4.2[Gpa]
Gzx	4.2[Gpa]
density	1.8 g/cc

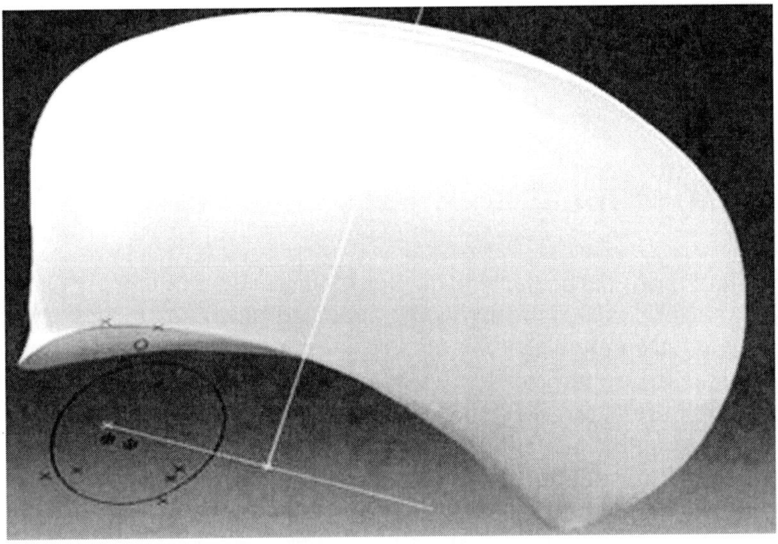

Figure 1. Single surface Model Propeller Blade

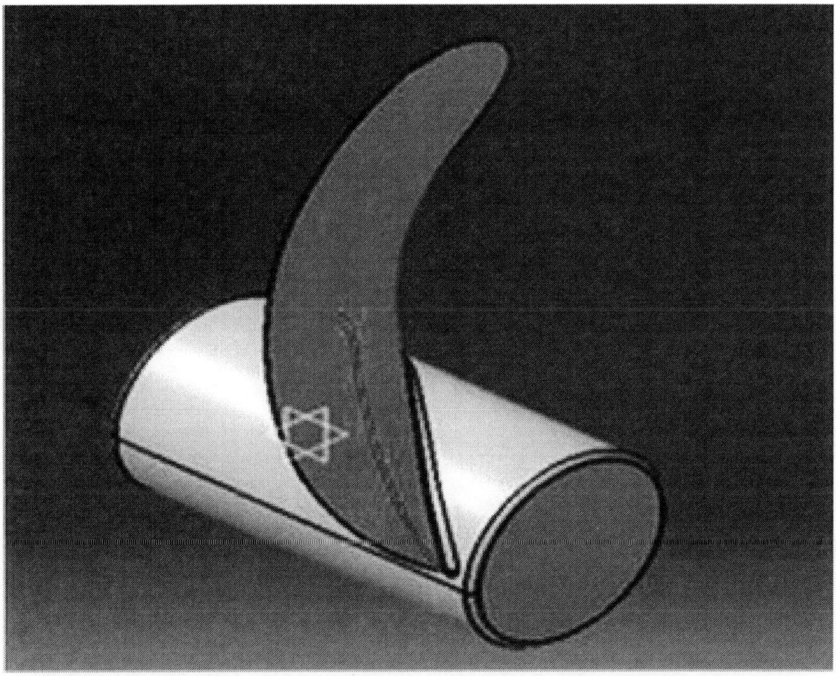

Figure 2. Solid Model of Single blade Propeller

For modelling of such complicated shapes of blades CatiaV5R20 was selected as base line modeller tool for analysis. The surfaces being important parameter for modelling the blade these surfaces are joined by a series of air foil points at various radii. The air foil points generated individually are rotated at respective pitch angles. All the rotated sections are projected onto the cylinder. These projection points are joined by means of a multi spline option to generate the surface of the blade. The solid model of the blade was generated using surface edit options. The air foil points and corresponding blade surface are shown in the following figures.

3.1. Propeller Blade Specifications

Table 3. Propeller Blade Characteristics

Type of series	wageningen B screw series propeller
Delivered power (P_D)	648KW
Advance speed (V_A)	4.372m/s
Propeller rate of rotation(N)	380 rpm
Propeller diameter (D)	2.12m
Number of blades (Z)	4
Blade area ratio (A_E/A_0)	0.70
P/D Ratio	0.9

Figure 3. Solid Model of 4 Bladed Propeller

3.2. Meshing of Propeller Blade

The solver used for developing fine mesh over the blade surface is Hyper mesh 11.0. Intially two dimensional meshes was generated for the blade surface using quadrilateral elements. Later the 2Dmesh was converted to three dimensional mesh using solid map options. The total numbers of finite elements created for the model are given below.

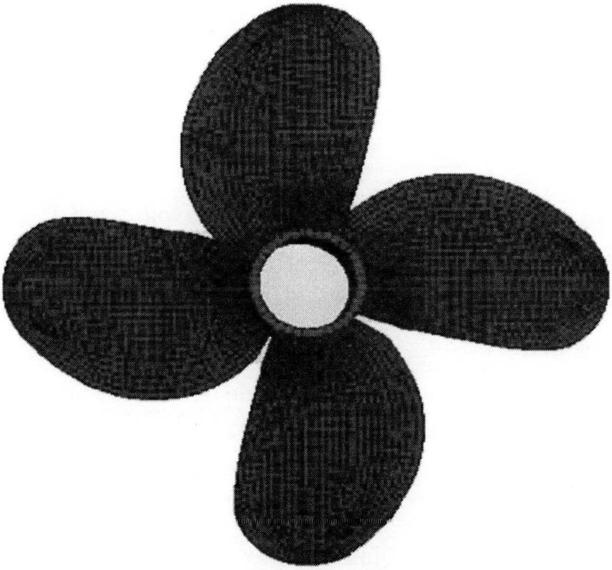

Figure 4. Meshed Model of Metallic Propeller Blade

Figure 5. Composite Propeller Blade

4. MATHEMATICAL MODELING OF TWISTED BLADE

Assumptions

The equation of motion is derived based on the following assumptions.

The rate of pre twist along the longitudinal axis of the blade is uniform.

The shear and rotary inertia effects of the blade are negligible due to slender shape of the blade.

The neutral and centroidal axis of the blade coincides with each other in the cross section.

Gyroscopic coupling between the stretching and bending motions are negligible and

No external forces act on the blade.

4.1. Geometry and Deformation of Shell Element

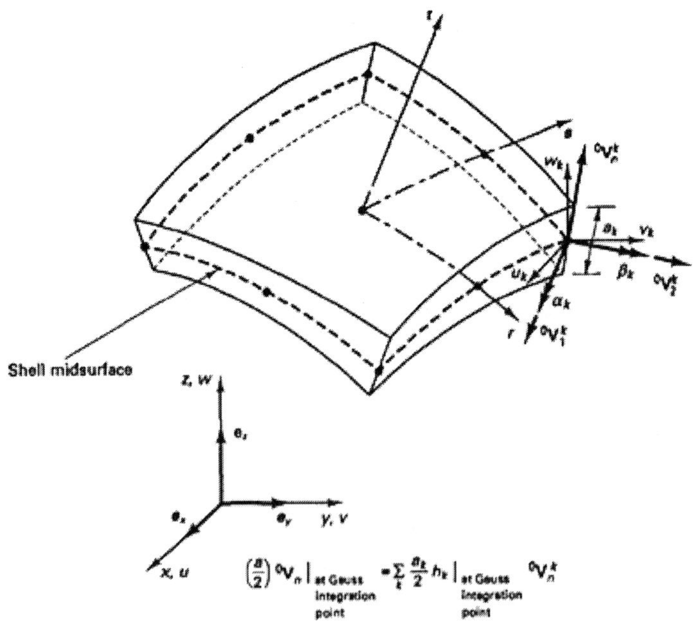

Figure 6. General shell element

$$\ell_x(r,s,t)=\sum_{k=1}^{q} h_k \ell_{xk}+\frac{t}{2}\sum_{k=1}^{q} a_k h_k\, {}^{e}V_{nk}^{k}$$

$$\ell_y(r,s,t)=\sum_{k=1}^{q} h_k \ell_{yk}+\frac{t}{2}\sum_{k=1}^{q} a_k h_k\, {}^{e}V_{ny}^{k} \qquad (1)$$

$$\ell_z(r,s,t)=\sum_{k=1}^{q} h_k \ell_{zk}+\frac{t}{2}\sum_{k=1}^{q} a_k h_k\, {}^{e}V_{nz}^{k}$$

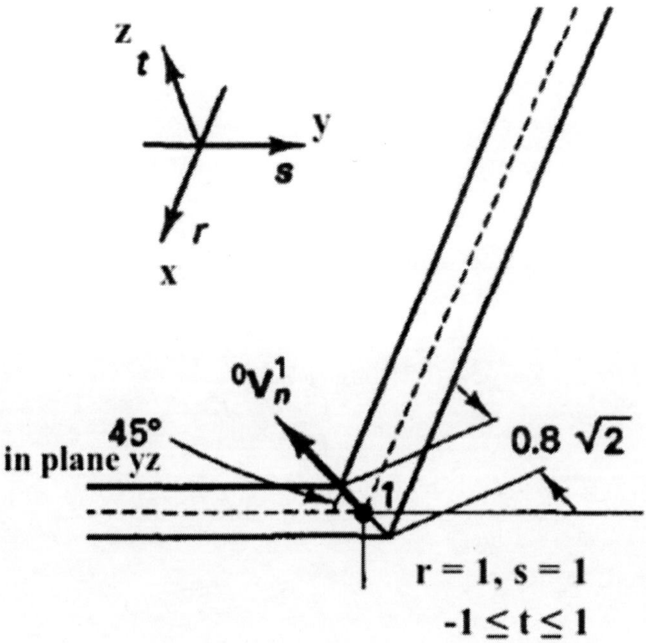

Figure 7. Coordinate Interpolation

$$\ell_x(r,s,t) = \sum_{k=1}^{q} h_k \ell_{xk} + \frac{t}{2} \sum_{k=1}^{q} a_k h_k \, {}^{e}V_{nk}^{k}$$

$$\ell_y(r,s,t) = \sum_{k=1}^{q} h_k \ell_{yk} + \frac{t}{2} \sum_{k=1}^{q} a_k h_k \, {}^{e}V_{ny}^{k} \qquad (2)$$

$$\ell_z(r,s,t) = \sum_{k=1}^{q} h_k \ell_{zk} + \frac{t}{2} \sum_{k=1}^{q} a_k h_k \, {}^{e}V_{nz}^{k}$$

$${}^{0}V_n^1 = \begin{bmatrix} 0 \\ \frac{-1}{\sqrt{2}} \\ \frac{1}{\sqrt{2}} \end{bmatrix} \qquad (3)$$

For t=1

$$x = x_1$$

$$y = y_1 + 0.5 * 0.8 \sqrt{0.2 * (\frac{-1}{\sqrt{2}})} \qquad (4)$$

$$z = z_1 + 0.5 * 0.8 \sqrt{0.2 * (\frac{-1}{\sqrt{2}})}$$

For t= -1

$$x = x_1$$
$$y = y_1 - 0.5*0.8\sqrt{0.2*(\frac{-1}{\sqrt{2}})} \qquad (5)$$
$$z = z_1 - 0.5*0.8\sqrt{0.2*(\frac{1}{\sqrt{2}})}$$

Displacement Interpolation

$$u_x(r,s,t) = \sum_{k=1}^{q} h_k u_k + \frac{t}{2}\sum_{k=1}^{q} a_k h_k \, {}^{e}V_{nk}^{k}$$
$$v_y(r,s,t) = \sum_{k=1}^{q} h_k v_k + \frac{t}{2}\sum_{k=1}^{q} a_k h_k \, {}^{e}V_{ny}^{k} \qquad (6)$$
$$w_z(r,s,t) = \sum_{k=1}^{q} h_k w_k + \frac{t}{2}\sum_{k=1}^{q} a_k h_k \, {}^{e}V_{nz}^{k}$$

Strain displacement matrix B(r,s,t)

$$\frac{\partial}{\partial x} = \frac{\partial}{\partial r}\frac{\partial r}{\partial x} + \frac{\partial}{\partial s}\frac{\partial s}{\partial x} \qquad (7)$$

J_{ij}^{-1} ↓

$$\begin{bmatrix} \frac{\partial u}{\partial r} \\ \frac{\partial u}{\partial s} \\ \frac{\partial u}{\partial t} \end{bmatrix} = \begin{bmatrix} \frac{\partial h_k}{\partial r}\begin{bmatrix} 1 & tg_{1x}^{k} & tg_{2x}^{k} \end{bmatrix} \\ \frac{\partial h_k}{\partial s}\begin{bmatrix} 1 & tg_{1x}^{k} & tg_{2x}^{k} \end{bmatrix} \\ h_k\begin{bmatrix} 0 & g_{1x}^{k} & g_{2x}^{k} \end{bmatrix} \end{bmatrix} \begin{bmatrix} u_k \\ \alpha_k \\ \beta_k \end{bmatrix} \qquad (8)$$

General Shell element

Stress strain law

$$\tau = C_{sh}\varepsilon \tag{9}$$

$$\tau^T = [\tau_{xx}\ \tau_{yy}\ \tau_{zz}\ \tau_{xy}\ \tau_{yz}\tau_{zx}] \tag{10}$$

$$\varepsilon^T = [\varepsilon_{xx}\ \varepsilon_{yy}\ \varepsilon_{zz}\ \varepsilon_{xy}\ \varepsilon_{yz}\ \varepsilon_{zx}] \tag{11}$$

C_{sh}

$$= Q_{sh}^T\left(\frac{E}{1-\nu^2}\begin{bmatrix} 1 & \nu & 0 & 0 & 0 & 0 \\ & 1 & 0 & 0 & 0 & 0 \\ & & 0 & 0 & 0 & 0 \\ symmetric & & & \frac{1-\nu}{2} & 0 & 0 \\ & & & & k\frac{1-\nu}{2} & 0 \\ & & & & & k\frac{1-\nu}{2} \end{bmatrix}\right)Q_{sh} \tag{12}$$

In which E and υ are the elastic and poisons ratio respectively. The material property matrix

$$[D] = [T_\varepsilon]^T [D'][T_\varepsilon] \tag{13}$$

where the transformation matrix $[T_\varepsilon]$ is

$$[T_{\varepsilon}] = \begin{bmatrix} l_{11}^2 & l_{12}^2 & l_{13}^2 & l_{11}l_{12} & l_{12}l_{13} & l_{13}l_{11} \\ l_{21}^2 & l_{22}^2 & l_{23}^2 & l_{21}l_{22} & l_{22}l_{23} & l_{23}l_{21} \\ l_{31}^2 & l_{32}^2 & l_{33}^2 & l_{31}l_{32} & l_{32}l_{33} & l_{33}l_{31} \\ 2l_{11}l_{21} & 2l_{12}l_{22} & 2l_{13}l_{23} & \begin{pmatrix} l_{11}l_{22} \\ +l_{21}l_{12} \end{pmatrix} & \begin{pmatrix} l_{12}l_{23} \\ +l_{22}l_{132} \end{pmatrix} & \begin{pmatrix} l_{13}l_{21} \\ +l_{23}l_{11} \end{pmatrix} \\ 2l_{21}l_{31} & 2l_{22}l_{32} & 2l_{23}l_{33} & \begin{pmatrix} l_{21}l_{32} \\ +l_{31}l_{22} \end{pmatrix} & \begin{pmatrix} l_{22}l_{33} \\ +l_{32}l_{23} \end{pmatrix} & \begin{pmatrix} l_{23}l_{31} \\ +l_{33}l_{21} \end{pmatrix} \\ 2l_{31}l_{11} & 2l_{32}l_{12} & 2l_{33}l_{13} & \begin{pmatrix} l_{31}l_{12} \\ +l_{11}l_{32} \end{pmatrix} & \begin{pmatrix} l_{32}l_{13} \\ +l_{12}l_{33} \end{pmatrix} & \begin{pmatrix} l_{33}l_{11} \\ +l_{13}l_{31} \end{pmatrix} \end{bmatrix} \quad (14)$$

Where l_{ij} are the directional cosines of the unit vector

$$\begin{aligned} l_1 &= \cos(e_x, e_{\bar{r}}); \\ l_2 &= \cos(e_x, e_{\bar{s}}); \\ l_3 &= \cos(e_x, e_t); \end{aligned} \quad (15)$$

$$\begin{aligned} m_1 &= \cos(e_x, e_{\bar{r}}); \\ m_2 &= \cos(e_x, e_{\bar{s}}); \\ m_3 &= \cos(e_x, e_t); \end{aligned} \quad (16)$$

$$\begin{aligned} n_1 &= \cos(e_x, e_{\bar{r}}); \\ n_2 &= \cos(e_x, e_{\bar{s}}); \\ n_3 &= \cos(e_x, e_t); \end{aligned} \quad (17)$$

Shear Locking

$$\varepsilon = \bar{\varepsilon}_{rr} g^r g^r + \bar{\varepsilon}_{ss} g^s g^s + \bar{\varepsilon}_{rt}(g^r g^t + g^t g^r)$$
$$+ \bar{\varepsilon}_{rs}(g^r g^s + g^s g^r) + \bar{\varepsilon}_{st}(g^s g^t + g^t g^s)$$
$$\downarrow \tag{18}$$
$$\underbrace{\bar{\varepsilon}_{rr} g^r g^r + \bar{\varepsilon}_{ss} g^s g^s + \bar{\varepsilon}_{rt}(g^r g^t + g^t g^r)}_{\text{In - layer Strains}}$$
$$\bar{\varepsilon}_{rs}(g^r g^s + g^s g^r) + \bar{\varepsilon}_{st}(g^s g^t + g^t g^s)$$

Transverse Shear strains

The element stiffness matrix is computed from

$$[K] = \int_{\Omega_e} [B]^T [D][B] d\Omega \tag{19}$$

And the element mass matrix is given as

$$[M] = \int_{\Omega_e} [N]^T [\rho][N] d\Omega \tag{20}$$

In general the equation of motion can be written

$$M\ddot{q} + C\dot{q} + kq = r \tag{21}$$

WhereM, C, *k*mass, damping, stiffness matrix

$\ddot{q}, \dot{q}, q$ acceleration, velocity, displacement

r external force vector

5. APPLICATION OF BOUNDARY CONDITIONS

For determining the natural frequencies Ansys14.0 is used as a postprocessor for metallic blade. Solid 45 & solid 46 element type are used for metallic and composite blades. The hub is constrained with all degrees of freedom arrested, assuming the propeller blade as a cantilever beam fixed at one end. The mode shapes and corresponding natural frequencies are determined. The composite material with various layup sequences are used to evaluate the best one that match up with isotropic material.

6. MODAL ANALYSIS OF PROPELLER BLADE

The modes of vibration of a propeller blade beyond the fundamental and first torsional/flexural modes are extremely complex. This complexity arises from the non-symmetrical outline of the blade. The variable thickness distribution both chordally and radially and the twist of the blade caused by changes in the radial distribution of pitch angle. Specifications for such rotating components with substantial dynamic loads often require the structural modes occurring in specified frequency range. The dynamic analysis of complex systems is often substantially complicated for such cases computer finite element methods are called for. A powerful Eigen extraction method for natural vibration analysis for rotating components is used. This method is based on a Block lancoz algorithm. Lancoz method has been developed for finding the some or all of the eigen values[12, 13] and eigen vectors of large symmetric sparse matrix. For dynamic analysis an unsymmetrical Block Lancoz algorithm may be employed to handle matrices with repeated eigen values, permit dynamic response analysis with multiple simultaneously applied loads/accommodate unsymmetrical damping matrices [14, 15]Lancoz method may also be used to reduce order structural models for component synthesis and control applications[16]. By applying the necessary boundary conditions and respective mechanical and material properties the natural frequencies and corresponding mode shapes of both metallic and composite finite element model are extracted. This analysis represents the un damped free vibrations of the propeller blade in absence of damping and applied loads.

7. RESULTS

In this study the free vibration characteristics of propeller blade with different ply ups and number of layers were determined. The comparison of natural frequencies and corresponding mode shapes were plotted. The numerical results from Table 4 and fig predict that, the first natural frequency of propeller blade when replaced with MAB has high operational frequencies and composite with first layup sequence produced the lowest. The percentage variation in terms of natural frequencies between MAB & composite with first layup is varied between 0.260% to 0.3684%. From Table 5 and fig shows that with change in number of layers from 4 to 8 enhances the composite material to match up with the frequencies of metallic one. From results it can be predicted that with further increase in number of layers and change of layup sequences the composite propeller blade is made to operate higher frequencies than that of metallic one.

Table 4. Composite with various stacking sequences

Stacking Sequence
$(45_2/0_4/90/45/0_2/-45/0/90)_{sym}$
$(0/45/-45)_7(0/90)_2$
$(45/-45/0)_7(0/90)_2$
$(45/0/-45)_7(0/90)_2$
$(-45/0/45)_7(0/90)_2$

Table 5. Natural Frequencies of composite for 4-Layers

MAB	SQ1	SQ2	SQ3	SQ4	SQ5
459.11	329.53	289.96	329.53	339.54	339.54
1511.3	1106.8	1064.5	1106.8	1120.0	1120.0
1725.5	1355.6	1217.0	1355.6	1355.6	1355.6
3166.7	2325.5	2390.2	2325.5	2348.2	2348.2
4217.7	3306.0	3014.7	3306.0	3335.3	3335.3
5846.5	4191.6	3978.9	4191.6	4325.4	4325.4
6382.8	4694.8	4600.6	4694.8	4888.5	4888.5
6888.8	4953.3	4962.2	4953.3	5091.8	5091.8
7755.2	5914.3	5489.1	5914.3	5988.2	5988.2
8752.4	6403.3	6240.7	6403.3	6472.4	6472.4

Table 6. Natural Frequencies of composite for 8-Layers

SQ1	SQ2	SQ3	SQ4	SQ5
333.96	338.22	336.02	319.40	326.38
1111.6	1148.3	1134.2	1119.0	1125.4
1362.4	1331.7	1397.4	1322.8	1351.0
2341.3	2497.8	2380.0	2414.6	2404.1
3337.4	3250.6	3365.7	3246.9	3304.0
4290.1	4400.1	4336.3	4250.8	4303.9
4856.5	4880.7	4926.4	4823.6	4879.4
5061.8	5281.0	5172.4	5129.7	5153.9
5980.9	5890.2	6026.3	5840.7	5935.8
6454.6	6636.4	6593.8	6538.3	6550.0

Table 7. Natural Frequencies of composite for 16-Layers

SQ1	SQ2	SQ3	SQ4	SQ5
324.12	318.77	331.15	325.94	325.45
1121.5	1119.4	1130.9	1126.8	1125.0
1334.3	1319.3	1362.8	1344.1	1345.7
2404.8	2418.0	2394.7	2406.9	2406.9
3276.2	3245.7	3333.3	3296.7	3296.2
4292.7	4246.2	4320.0	4292.8	4295.5
4872.2	4820.3	4902.5	4870.4	4870.9
5153.6	5126.3	5157.3	5142.2	5147.9
5896.2	5846.6	5979.3	5924.9	5924.0
6542.5	6536.5	6581.7	6567.6	6552.7

Table 8. Natural Frequencies of composite for 25-Layers

SQ1	SQ2	SQ3	SQ4	SQ5
322.84	320.60	328.90	325.94	325.52
1116.0	1115.7	1123.8	1126.8	1120.1
1313.4	1319.3	1346.4	1344.1	1335.1
2410.8	2406.4	2391.7	2406.9	2399.3
3232.1	3247.8	3305.0	3296.7	3279.8
4275.7	4246.5	4294.2	4292.8	4278.0
4839.7	4819.9	4872.4	4870.4	4250.5
5138.7	5080.2	5101.3	5142.2	5093.4
5831.1	5852.1	5940.8	5924.9	5901.8
6517.8	6497.5	6529.3	6567.6	6512.3

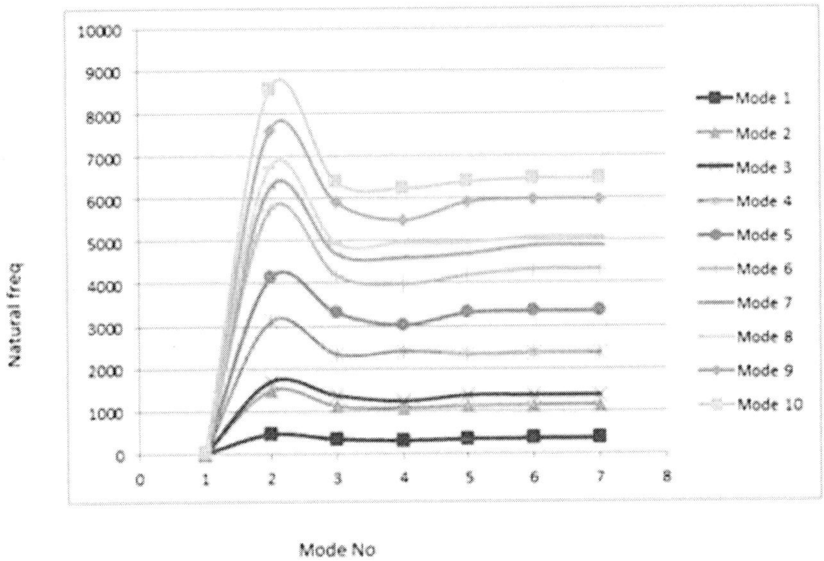

Figure 8. NF s Mode No (4-Layers)

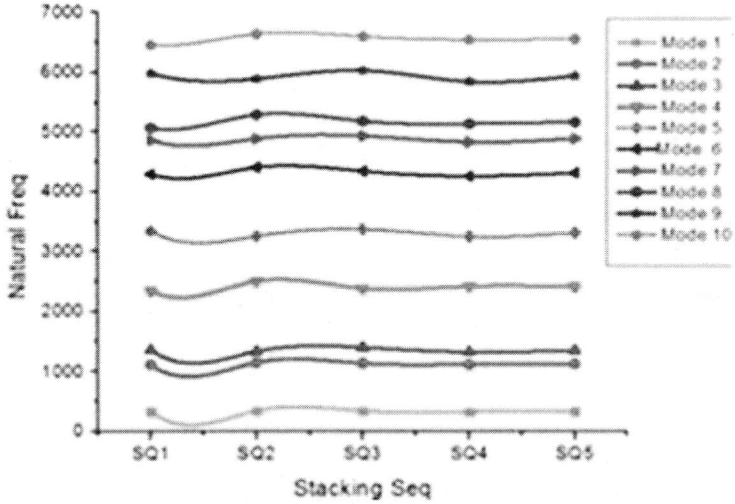

Figure 9. NF s Mode No (8-Layers)

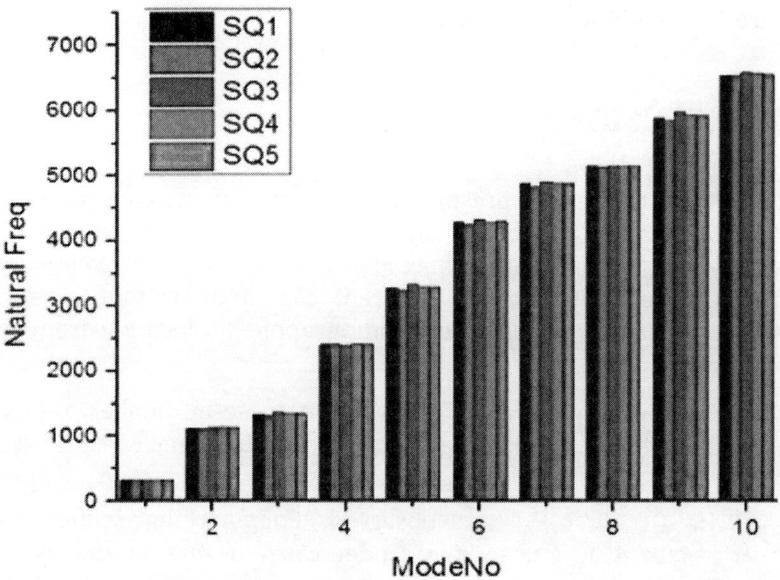

Figure 10. NF s Mode No (16-Layers)

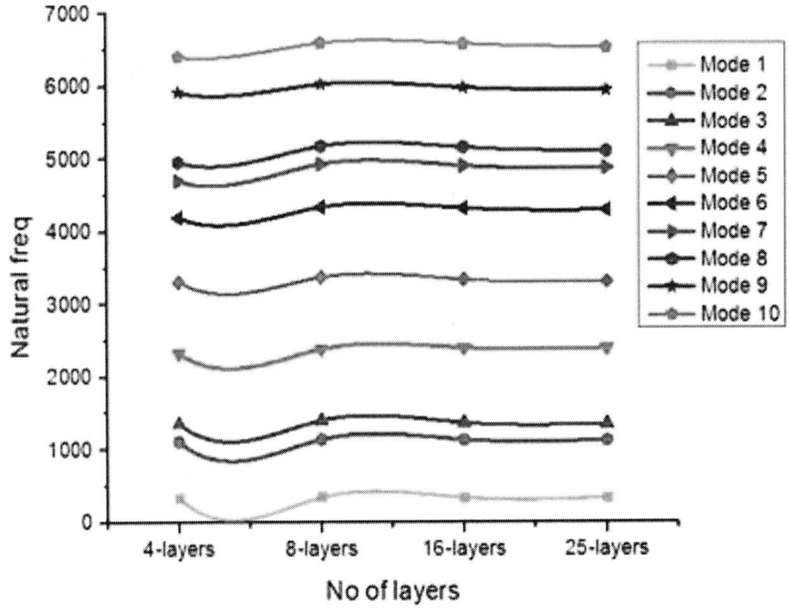

Figure 11. NF s Mode No (25-Layers)

8. CONCLUSIONS

1. From the results presented it is clear that the change in fibre orientation, number of layers and laminate stacking sequence yield to different dynamic behaviour of the component. Different layers apply different contributions to the overall stiffness of the blade depending upon the location from mid plane.
2. From Table 4 & Table 5 with the increase in number of layers from 4 to 8 an increase in the flexural frequencies and torsion frequencies for stacking sequence of $(\pm45/0)_{11}$, $(45/-45/0)_{11}$, $(0/\pm45)$ are observed. For stacking sequence of $(45/0/-45)_{11}$& $(-45/0/45)_{11}$ decrease in torsion modes and flexural are observed.
3. From Table 5 & Table 6 the increase in number of layers from 8 to 16 varied in steps increase in the first bending frequency and torsion frequencies for stacking sequence of $(\pm45/0/90)$ are observed. For stacking sequence of $(\pm45/0)_{11}(0/90)_7$, $(0/\pm$

45)$_{11}$(0/90)$_7$ decrease in torsion and flexural modes are observed, while for (+45/0/-45)$_7$(0/90)$_2$, (-45/0/+45)$_7$(0/90)$_2$ increase in both the flexural and torsion are observed.

4. From Table 6 & Table 7 it is seen that with further increase in number of layers from 16 to 25 layers decrease in flexural and torsional frequencies are observed in all the stacking sequences.

5. From Table 4, Table 5, Table 6 the stacking sequence (± 45/0)$_7$ (0/90)$_2$for composite has got highest bending frequencies and torsional frequency. From the results it is possible to verify the influence of stacking sequence of the laminate on torsional vibration the laminate with (± 45°) has larger torsional natural frequencies. The natural frequencies related to stiffness of the structure (±45) are much stiffer in torsion. The laminate (0/45) of fibres has lowest torsional frequency than the other lamination schemes and most of the fibres are oriented at 0° direction and thus appropriate for flexural modes (bending).In fact this can be applied that the fibres oriented at 0° are appropriate for flexural loads and fibres oriented at 45° area more appropriate for torsion loads. In order to attain increase in bending & torsion modes an optimum numbers of layers are selected in between 8 and 16 layers. From the results it can be predicted that the stacking sequence number 3 has got highest operating frequencies for 16 layers close to the metallic one. The desired vibration characteristics can be attained for the blade without modifying the shape. Hence the ply orientation, material selection should be optimized to attain maximum operating frequencies.

REFERENCES

1. J.E.Conolly, "Strength Of Propellers", reads in London at a meeting of the royal intuition of naval architects on Dec 1,1960,pp 139-160.

2. Leissa, K.I. Jacob, Three-dimensional vibrations of twistedcantilevered parallelepipeds, Journal of Applied Mechanics-Transactions of the ASME 53, (1986), 614-618.

3. R.E. Kielb, A.W. Leissa, J.C. MacBain, Vibrations of twistedcantilever plates-a comparison of theoretical results, International Journal for Numerical Methods in Engineering 21, (1985), 1365-1380.

4. South well. R and F. Gough, The free transverse vibration of airscrew blade, British A.R.C. Reports and Memoranda, 1921, No. 766.

5. P.P. Fried mann, F. Straub, Application of the finite element methodto rotary-wing aero elasticity, Journal of American Helicopter Society 25 (1980) 36-44.

6. [6] N.T. Sivaneri, I. Chopra, Dynamic stability of a rotor blade usingfinite element analysis, AIAA Journal 20 (1982) 716-723.

7. [7] Decolon.Estimation a posteriori et adaptation de malliages. coordinateur Michel Fortin, Hermes, 2000.

8. M.R.M. Crespo Da Silva, A comprehensive analysis of thedynamics of a helicopter rotor blade, International Journal of Solids and Structures 35 (7–8) (1998) 619-635.

9. V. Ramamurti, R. Kielb, Natural frequencies of twisted rotating plates, Journal of Sound and Vibration 97 (1984) 429-449.

10. S. Sreenivasamurthy, V. Ramamurti, A parametric study ofvibration of rotating pre-twisted and tapered low aspect ratiocantilever plates, Journal of Sound and Vibration 76 (1981) 311-328.

11. A.W. Leissa, J.K. Lee, A.J. Wang, Rotating blade vibration analysisusing shells, Journal of Engineering for Power, Transactions of American Society of Mechanical Engineers 104 (1982) 296-302.

12. T. Tsuiji, M. Shugyo, T. Yamashita, Free vibration analysis ofrotating thin twisted plates, Transactions of Japan Society of Mechanical Engineers 61 (C) (1995) 4575-4586 (in Japanese).

13. X.X. Hu, T. Tsuiji, Free vibration analysis of rotating twistedcylindrical thin panels, Journal of Sound and Vibration 222. 1999 209-224.

14. Lo,H.,A nonlinear problem in the bending vibration of a rotatingbeam, J.Appl.Mech., Trans.ASME, 1952, 19, p.461.

15. Boyce, W.e. Diprima, R.C. and andleman, G.H., Vibrations of rotating beams of constant sections,proc.and Us Natl.cong.Appl,Mech.,ASME, 1954,p.165.

16. Schilhansl, M.J., Bending frequency of arotatingcantileverbeam, J.Appl.Mech., ASME, 25, 1958, p. 28.

17. Hirsch,G.,Investigation of vibration in bending of rotating turbineblades on non-rigid support, 1958, Jahrbuck Wissens chart., GessellschLuftahrt p. 174.

18. HorwayG.,chord wise and beam wise bending of hinged rotorblades, J Aero. Sci., 1948, 15, p.497.

19. Niordson, F., Vibration of turbine blades with loose hingesupports, 1954, Engr.Digest, 15, p. 359.

20. Garland C.F.,The normal modes and vibrations of beams havingnoncollinearelasticandmassaxes,J.appl.Mech.,Trans.ASME , 1940,p.a-97.

21. Kapuria S, Alam N., Efficient layerwise finite elementmodel for dynamic analysis of laminated piezoelectric beams, Computer Methods in Applied Mechanics and engineering, 2006, Vol. 195, pp. 2742-2760.

22. Yõldõrõm V, Kõral E., Investigation of the rotary inertia and sheardeformation effects on the out-of-plane bending and torsionalnatural frequencies of laminated beams, Composite Structures, Vol. 49, 2000, pp. 313-320.

23. Khdeir A.A., Reddy J.N., Free vibration of cross-ply laminatedbeams with arbitrary boundary conditions, International Journal of Engineering Science, 1994, Vol. 32, pp. 19711980.

24. Crawley, E. F. The Natural Modes of Graphite/Epoxy CantileverPlates and Shells, Journal of Composite Materials, 1979, Vol. 13, pp. 195-205.

25. Wang, J. T. S., Shaw, D. and Mahrenholtz, O. Vibration of RotatingRectangular Plates, Journal of Sound and Vibration, Vol. 112, No. 3, 1987, pp. 455-468.

26. Shaw, D., Shen, K. Y. and Wang, J. T. S. Flexural Vibration of Rotating Rectangular Plates of Variable Thickness, Journal of Sound and Vibration, 1988 Vol. 126, No. 3, pp. 373-385.

27. Qatu M. S. and Leissa A. W. Vibration studies for Laminated Composite Twisted Cantilever Plates, International Journal of Mechanical Sciences, 1991, Vol.33, pp. 927-940.

28. BirG.S, and Migliore P.G. Computerized method for preliminary structural design of composite wind turbine blades [J]. Journal of solar energy engineering, 2001, vol. 123, iss. 4, pp. 372-381.

29. Alejandro D. Otero, and Fernando L. Ponta. Structural analysis of wind turbine blades by a generalized Timoshenko beam model [J]. Journal of Solar Energy Engineering, 2010, vol. 132, pp. 151-158.

30. Liu Wang-yu, Zhang Hai-quan, and Zeng lin. Reliability analysis of bionic wind turbine blades based on the response surface methodology [J]. Journal of Solar Energy Engineering., 2010, vol. 31, iss. 9, pp. 1204-1208.

31. Shakeri M, Yas MH, Gol MG, Optimal stacking sequence of laminated cylindrical shells using genetic algorithm, MechAdv Mater Struct, 2005, 305-312.

8

CHAPTER

Investigation on Hydrodynamic Performance of a Marine Propeller in Oblique Flow by RANS Computations

Jianxi Yao

Technical University of Berlin, Germany

ABSTRACT

This paper presents a numerical study on investigating on hydrodynamic characteristics of a marine propeller in oblique flow. The study is achieved by RANS simulations on an open source platform - OpenFOAM. A sliding grid approach is applied to compute the rotating motion of the propeller. Total force and moment acting on blades, as well as average force distributions in one revolution on propeller disk, are obtained for 70 cases of com- binations of advance ratios and oblique angles. The computed results are compared with available experimental data and discussed.

Keywords: Propeller; Hydrodynamic performance; Oblique flow; RANS simulation; OpenFOAM.

INTRODUCTION

Numerous studies based on experiments or computations have been carried out to investigate propeller open water characteristics which are main ways to access propeller hydrodynamic performance in open water currently. Most studies only considered the case of propeller in straight ahead flow. However, under real operating conditions working propeller behind ship usually locates in so complicated wake that the propeller shows quite different hydrodynamic performance. This is mainly due to the disturbance of ship hull to inflow before propeller, which results in each blade angle of attack of propeller becomes different and the loads acting on blades in axial direction is not symmetric anymore. So that it is necessary to study the cases of propellers in complicated flow, so as to obtain a fuller knowledge and understanding of propeller hydrodynamic performance.

On the other hand, in recent years numerical methods based on solving Reynolds-Averaged Navier-Stokes (RANS) equations have been made remarkable progress and widely applied to address problems in ship hydrodynamics, such as manoeuvre. For instance, Cura-Hochbaum (2006) predicted ship manoeuvring behaviours based on the determination of manoeuvring derivatives by means of virtual PMM (Planar Motion Mechanism) tests. A similar study was also done by Simonsen et al. (2012) who performed the standard deep water IMO (International Maritime Organization) manoeuvring simulations for a container ship model with the hydrodynamic coefficients obtained from both RANS computations and measurements. In their simulations, effects of real working propeller-on flow were approximated by either a simple prescribed body force model or a model based on a database consisting of force distributions on propeller disk for a range of advance ratios, as using real discretized propeller usually requires small length of time step, making the simulations extremely time-consuming. The experience from the first workshop on verification and validation of ship manoeuvring simulation methods-SIMMAN (Stern et al., 2008) has shown that present body force models used are required to be improved for highly accurate simulations. Therefore, in order to develop more sophisticated body force models insightful knowledge of force distributions on propeller disk is needed when propeller rotates in complicated flow.

On above points of view, an investigation on hydrodynamic characteristics of a marine propeller in oblique flow is made in this study. Only few authors have reported on this aspect until now. Dubbioso et al., 2013 and Dubbioso et al., 2014 analyzed performance of a marine propeller in oblique flow by RANS computations. In their work the hydrodynamic forces acting on blades and flow features around the propeller were obtained and discussed in details. Unfortunately the simulations were not validated due to lack of experimental data. Krasilnikov et al. (2009) used an unsteady RANS method to investigate on propeller blade forces for a podded propeller operating in pulling and pushing modes in oblique flow. They found by the analyses that blades of pulling propeller experience comparable amplitudes and load levels at positive and negative heading angles being mainly affected by the cross flow. Shamsi and Ghassemi (2013)studied the effects of oblique angles on characteristics of podded propulsions. A RANS method with Moving Reference Frame (MRF) was used in their study. The predicted hydrodynamic forces showed good agreements with experimental data.

The present work is achieved by using the unsteady RANS solver in an open source Computational Fluid Dynamics (CFD) toolbox named OpenFOAM to simulate the cases of combinations of advance ratios and oblique angles. Seventy combinations are considered. During the simulations, Menter k-œ SST model (Menter and Kuntz, 2003) without wall function is used to model turbulence and an approach of sliding grid is applied to compute rotating motions of the propeller. Total force and moment acting on blades, as well as average force distributions in one revolution on propeller disk, are obtained. The computed results are compared with available experimental data and discussed.

OUTLINE OF METHOD

Mathematical équations

A right-handed coordinate system is used to describe the rotating motion of the propeller, where the origin *o* locates at the centre of the propeller, *x*-axis towards cap as depicted inFig. 1 and *z*-axis vertically downward. Since the rotating motion in three-space is symmetric with respect to x-axis, the oblique angle *a* could be generally defined as the angle from x-axis to the direction of free-stream flow. It is clear that the range of *a* is from 0° to 180°. Here *a* is defined in *o-xy* plane without loss of generality.

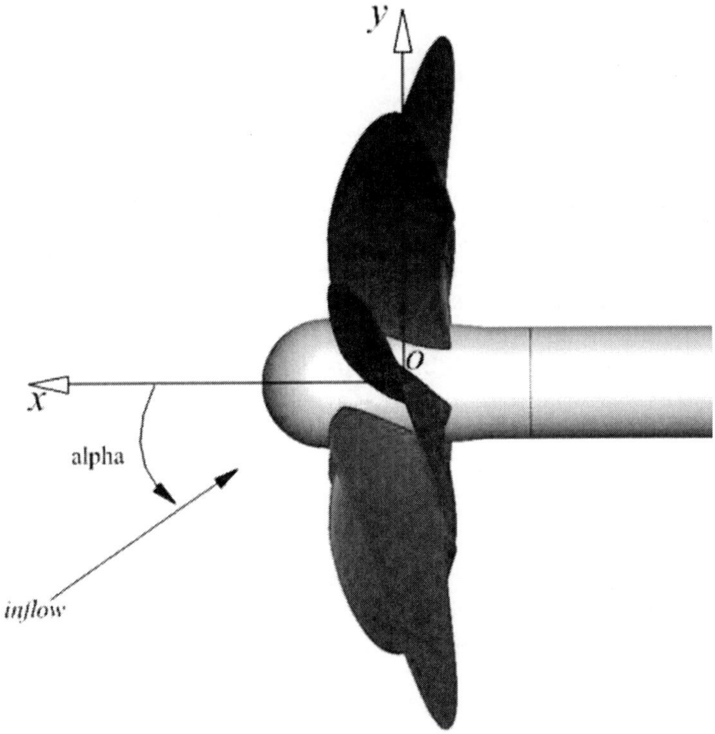

Figure 1. Definitions of coordinate system and oblique angle (top view).

Under the assumption of incompressible Newtonian fluid the flow around the propeller has to satisfy conservation equations of mass (continuity equation) and momentum (RANS equations), which could be written in tensor notation as

$$\frac{\partial U_i}{\partial x_i} = 0 \tag{1}$$

$$\frac{\partial (U_i)}{\partial t} + \frac{\partial (U_i U_j)}{\partial x_j} = f_i - \frac{1}{\rho}\frac{\partial p}{\partial x_i} + \frac{\partial}{\partial x_j}\left[\nu \left(\frac{\partial U_i}{\partial x_j} + \frac{\partial U_j}{\partial x_i} \right) + \tau_{ij} \right] \tag{2}$$

where x_t = (x, y, z) are independent coordinates, similarly *Ui* = (U, *V,, W)* are Reynolds-Averaged velocity components, *p* is pressure, *p* is water density, *V* is kinematic viscosity coefficient, f_i are body force components and *τij*is Reynolds stress tensor resulting from the time-averaged procedure to N-S (Navier-Stokes) equations. The Reynolds stress tensor is expressed under Bousinesq's eddy viscosity hypothesis as

$$\tau_{ij} = \nu_t \left(\frac{\partial U_i}{\partial x_j} + \frac{\partial U_j}{\partial x_i} \right) - \frac{2}{3}\delta_{ij} k \qquad (3)$$

where $\boldsymbol{V_t}$ is eddy viscosity, *δij,* is Kronecker symbol and *k* is turbulent kinetic energy. The body force*fi* are zero in present simulations.

The Menter k-œ SST model (Menter, 2003) without wall fonction is applied to approximate the eddy viscosity in équation (3) where SST is the acronym of Shear-Stress-Transport and m is specific dissipation rate. The transport equations are as follows.

$$\frac{\partial (k)}{\partial t} = \frac{\partial (U_i k)}{\partial x_i} = \tilde{P}_k - \beta^* k\omega + \frac{\partial}{\partial x_i}\left[(\nu + \sigma_k \nu_t) \frac{\partial k}{\partial x_i} \right] \qquad (4)$$

$$\frac{\partial (\omega)}{\partial t} + \frac{\partial (U_i \omega)}{\partial x_i} = \frac{\alpha P_k}{\nu_t} - \beta \omega^2 + \frac{\partial}{\partial x_i}\left[(\nu + \sigma_\omega \nu_t) \frac{\partial \omega}{\partial x_i} \right] + 2(1 - F_1)\sigma_{\omega 2} \frac{1}{\omega} \frac{\partial k}{\partial x_i} \frac{\partial \omega}{\partial x_i}. \qquad (5)$$

here the functions F_1 and F_2 are defined by

$$F_1 = \tanh\left\{ \left\{ \min\left[\max\left(\frac{\sqrt{k}}{\beta^* \omega y_w}, \frac{500\nu}{y_w^2 \omega} \right), \frac{4\rho\sigma_{\omega 2} k}{CD_{k\omega} y_w^2} \right] \right\}^4 \right\},$$

$$F_2 = \tanh\left[\left[\max\left(\frac{2\sqrt{k}}{\beta^{*}\omega y_w}, \frac{500v}{y_w^2\omega}\right)\right]^2\right]$$

with

$$CD_{k\omega} = \max\left(2\rho\sigma_{\omega 2}\frac{1}{\omega}\frac{\partial k}{\partial x_i}\frac{\partial \omega}{\partial x_i}, 10^{-20}\right)$$

and y_w the distance to wall. The turbulent eddy viscosity is defined by

$$v_t = \frac{a_1 k}{\max\left(a_1\omega, SF_2\right)}, \tag{6}$$

where

$$S = \sqrt{2S_{ij}S_{ij}}, \quad S_{ij} = \frac{1}{2}\left(\frac{\partial U_i}{\partial x_j} + \frac{\partial U_j}{\partial x_i}\right).$$

S, is the strain rate tensor and $a_1 = 0.31$ is a constant. And the production of turbulent kinetic energy is

equation)

$$P_k = \tau_{ij}\frac{\partial U_i}{\partial x_j} \rightarrow \tilde{P}_k = \min\left(P_k, 10 \cdot \beta^{*}k\omega\right)$$

All constants are computed by a blend from the corresponding constants via $c = c_1 F_1 + c_2 (1 - F_1)$. The values of model constants are specified to the default values given in OpenFOAM, which are: /3* = 0.09 , a = 0.553 , P_x= 0.075 , a_2 = 0.44, = 0.0828, < r_{k1} =0.85, u_{ml}= 0.5, a_{k2} = 1, cr_{m2} = 0.856.

Numerical method

The RANS solver in OpenFOAM (2014) is based on a finite volume technique which permits use of arbitrary polyhedral grids including hexahedron, tetrahedron, prism, and so on. A suite of basic discretization schemes and solution algorithms are available. In present applications, a second order upwind difference scheme and a central difference scheme are selected to approximate convective terms and diffusive terms respectively. Turbulence is discretized with a second order upwind difference. The PIMPLE algorithm which merges PISO (Pressure Implicit with Splitting of Operators) and SIMPLE (Semi-Implicit Method for Pressure-Linked Equations) algorithms is employed to couple the mass and momentum conversation equations. The systems of linear equation resulting from the discretized equations are solved by using the iterative solvers, Gauss-Seidel or generalized geometric multi-grid (GAMG).

Sliding grid approach

The capability in dealing with sliding grid is available in OpenFOAM. This technique is quite useful for simulations involving rotating parts e.g. rotating propeller. With the sliding grid approach the computational domain used may be divided into two parts: an outer static part and an inner dynamic part. The inner part is cylinder-shaped and could rotate relatively to the outer part. The two parts are connected by an interpolation scheme by which flow information is transformed between relative sliding interfaces. Fig. 2 shows partial view of an unstructured grid in a cut plane. As seen the inner part includes the geometry of a propeller and grid elements in it rotate during simulations. The red circle in the figure represents sliding interfaces on which the interpolation scheme is imposed via setting a boundary condition named Arbitrary Mesh Interface (AMI) in OpenFOAM.

GEOMETRY, GRID, BOUNDARY CONDITION AND GRID, TIME-STEP DEPENDENCY STUDIES

A propeller having 6 blades is taken as an example. The geometry of the propeller is shown in Fig. 3 and main particulars at full scale are listed in Table *1*. A model with scale 39.885 has been tested in the cavitation tunnel at TUB (Technical University of Berlin, Germany).

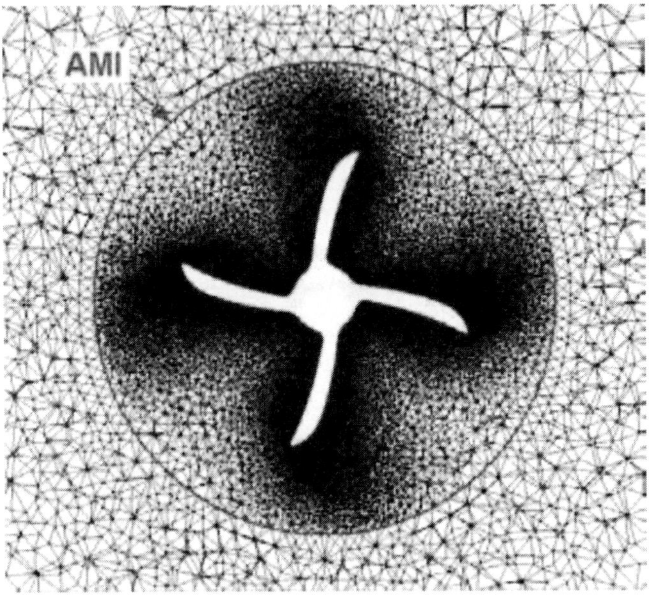

Figure 2. Sliding grid in a cut plane.

Figure 3. Geometry of the propeller. Fig. 4 Computational domain and boundary conditions.

Table 1. Main particulars of the propeller.

Diameter	8.3 *m*
Pitch ratio (mean)	1.03
Expanded area ratio	0.8034
Number of blade	6
Hub ratio	0.18
Direction of rotation (looking upstream)	Right-handed

According to the experiences from the 2nd International Symposium on Marine Propulsions , a rational computational domain is chosen for the computations, which is limited in a box with $-7.0D<x<3D$, $-5.0D<y<5.0D$ and $-5.0D<z<5.0D$. Fig. 4 shows the computational domain and boundary conditions (BC). BCs are summarized as follows.

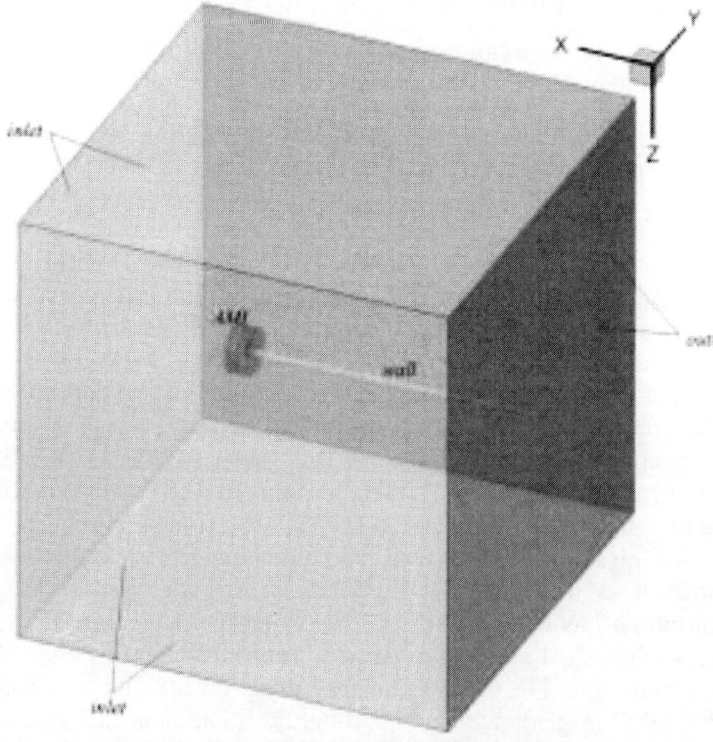

Fig. 4. Computational domain and boundary conditions.

- Inlet BC: at boundaries x = 3D, z = ± 5D and y = -5D, the free-stream values of velocity, *k*and *m* are specified, while zero gradient is given to pressure. The free-stream values of *k*and *m* could be estimated by following formulas.

$$k_{in} = \frac{3}{2}\left(I_t u_{in}\right)^2, \quad \omega_{in} = \frac{k_{in}^{1/2}}{C_\mu^{1/4} l_s}$$

where I_t in range of 1% ~ 10% is the turbulent intensity as a fraction of mean flow velocity u_n, *C^* = 0.09 is a constant and l_s is the turbulent length scale often close to the cell size.

- Outlet BC: at boundaries x = -7D and *y* = 5D, specify pressure and normal gradients of flow velocity, k, *m* to 0.
- Wall BC: on blades, cap and shaft no-slip condition is set, i.e. flow velocity and gradients of pressure, k, *m* are specified to 0.
- AMI BC: set AMI BC at the interfaces connecting the static and dynamic parts.

Due to the complicated configuration of the propeller a hybrid grid is employed. An unstructured grid is generated for the inner part, while multi-block structured grid for the outer part. The surface grid on blades consists of triangular elements and three mesh layers are attached at blade surfaces. The grid in the inner part is composed of triangular prisms in the three mesh layers and tetrahedrons in remaining space. The advantage of the mesh layers is that it is convenient to adjust spacing of first grid point to wall, but avoid strong increase of cell number. However, a special attention should be paid to the dimensionless distance to wall *(y +)* when generating grid since no wall function is used in this study. According to amount of pre-computations it has been found that specfying the spacing of first grid point to wall to 0.1 *mm* could get an appropriate y^+ (around 1) in present simulations. Fig. 6 shows surface grids on a blade of a coarse, medium and fine grid, which has roughly 1.5 *million,* 2.6 *million* and 5.2 *million* cells respectively. The grid in a mid-cut plane of the coarse grid is shown in Fig. 5.

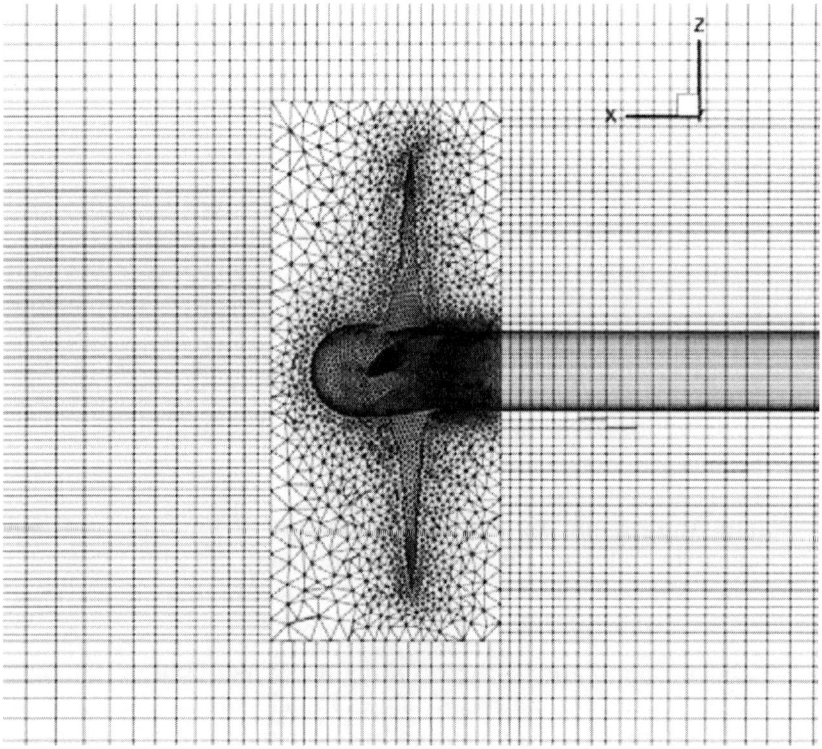

Figure 5. Grid in a mid-cut plane of the coarse grid.

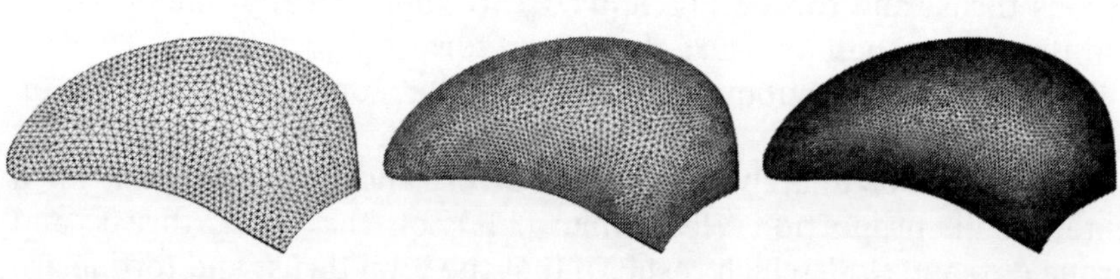

Figure 6. Surface grids on a blade: coarse (left), medium (middle), fine (right).

The simulation accuracy not only depends on grid resolution but also the length of time step used. In particular for cases of considering dynamic grid, time-step length would also affect significantly on numerical stability. For present cases it has been found that performing 50 time steps in one propeller revolution could ensure numerical stability and become instable using 25 time steps. During the simulations in each time step it applies 10 outer loop iterations, and in each iteration pressure is cor- rected twice. Note that if using smaller length of time step less number of iterations and pressure correction may also ensure accuracy.

Grid and time-step dependency studies are usually necessary for reliable numerical results. For this purpose, computations are conducted on the coarse, medium and fine grid respectively for a selected case of n=50 *rps*, u_0= 4.162 *m/s* and a=0° which correspond to J=0.4 and a Reynolds number R_e= 1.24*10[6], where *n* is revolution rate, *U0* flow velocity, and *J* advance ratio. The present Reynolds number is defined by $MWR \times D\sqrt{(0.7\pi nD)^2 + u_0^2} / \nu$ where *MWR=0.2565* is mean blade width ratio. In addition, for time-step dependency study computations are carried out on the medium grid for the same case, but performing different Number of time Step (NS) in one propeller revolution. The computations for grid dependency study perform 400 time steps in one propeller revolution, and for the time-step dependency study it keeps everything unchanged but only varies NS which is specified to 50, 100, 200, 400 and 800 here.

For all above computations, the resulting maximum Courant numbers are between 13 and 60, but mean values are all less than 0.1. It seems the present length of time step and cell size can maintain computational stability. The computed thrust and torque acting on blades are summarized in Table 2. It is seen from the grid dependency study that viscous thrust and torque (T_{vis} and Q_{vis}) are more greatly influenced by grid revolution than pressure thrust and torque (T_{pre} and Q_{pre}). There is still a discrepancy of about 10% between the computed $T_{vs}s$ based on the medium and fine grids, whereas only 0.6% between $T_{pre}s$. Fortunately the viscous thrust and torque account for a quite small proportion (less than 1%) of the total thrust and torque (T_{Ott} and Q_{tot}), which result in that the total thrust and torque are not sensitive to grid revolution.

In the time-step dependency study, it is found that as increase of NS the pressure thrust and torque change smaller and smaller, while the vicious thrust and torque are nearly unchanged. When NS reaches 200, the changes of total thrust and torque are less than 1%.

In order to illustrate clearly, the thrust convergence curves obtained from the computations of grid and time-step dependency studies are shown in Fig. 7. It is obviously concluded that all computations reach steady-state after 3 propeller revolutions.

Table 2. Summary of thrust and torque obtained from the grid and time-step dependency studies for the case of $J = 0.4$ and $a = 0°$.

Grid	NS	Thrust [*N*]				Torque [*Nm*]			
		T_{pre}	T_{vis}	T_{tot}	Err (%)	Q_{pre}	Q_{vis}	Q_{tot}	Err (%)
Grid dependency study									
Coarse	400	1756.34	-13.67	1742.67	**0.85**	-56.24	-2.00	-58.24	**0.50**
Medium	400	1773.02	-15.36	1757.66	**0.51**	-56.32	-2.21	-58.53	**0.15**
Fine	400	1783.76	-17.08	1766.68	---	-56.13	-2.49	-58.62	---
Time-step dependency study									
Medium	50	1570.71	-15.89	1554.82	**8.65**	-50.48	-2.25	-52.73	**7.36**
Medium	100	1717.47	-15.42	1702.05	**2.43**	-54.70	-2.22	-56.92	**2.12**
Medium	200	1759.71	-15.34	1744.37	**0.76**	-55.94	-2.21	-58.15	**0.65**
Medium	400	1773.02	-15.36	1757.66	**0.23**	-56.32	-2.21	-58.53	**0.20**
Medium	800	1777.20	-15.41	1761.79	---	-56.43	-2.22	-58.65	---

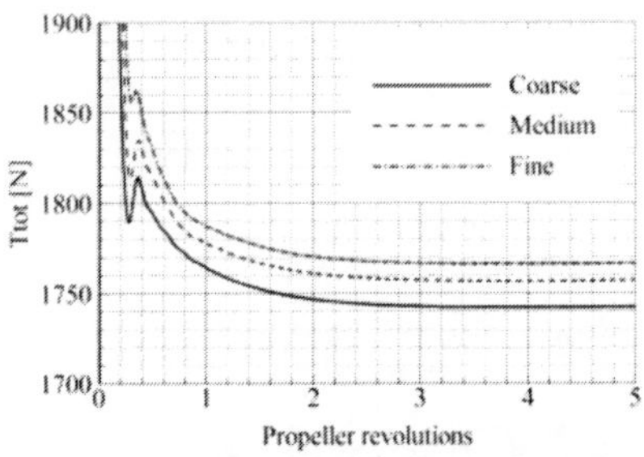

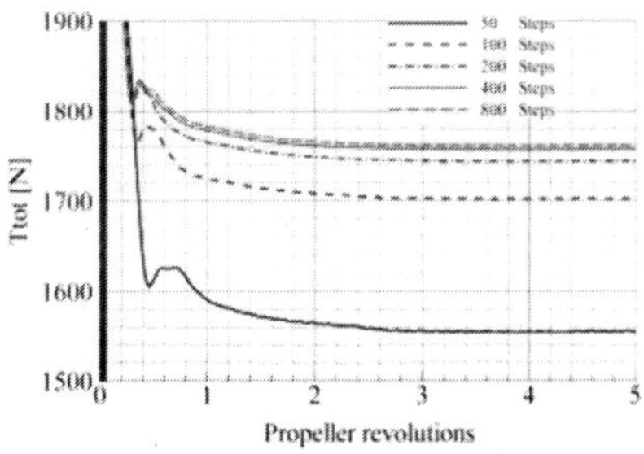

Figure 7. Comparison among thrust convergence curves obtained from grid (left) and time-step (right) dependency studies for the case of $J = 0.4$ and $a = 0°$.

RESULT AND DISCUSSIONS

Seventy cases of combinations of Js and as are considered, where J is varied from 0.1 to 1.0 with an interval of 0.1 and a from 0° to 30° with an interval of 5°. For all computations the magnitude of inflow velocity remains unchanged (here 4.162 *m/s)* and propeller revolution rate is determined for desired Js according to $J=u_0/(nD)$. The Reynolds numbers of these cases are between $0.537*10^6$ and $1.24*10^6$. In consideration of large amount of cases, all computations are carried out on the medium grid and perform 200 time steps in one propeller revolution. Based on the grid and time-step dependency studies in previous section, it could be expected that the discrepancy between the present numerical results and the results obtained by using the fine grid and larger number of time step should be within 2%.

In order to obtain average force distributions on propeller disk in one propeller revolution for all cases, each computation performs 4 revolutions and the pressure force and viscous stress acting on grid elements on one blade are written out every two time steps in the fourth revolution. This means there are 100 outputs for each computation. Then for each case the 100 force distributions on the blade are transformed onto propeller disk via an area-weighted interpolation scheme as sketched in Fig. 8, and averaged in the revolution. Note that only the force on one blade is outputted but not on all blades, as each blade experiences same loads at same phase in the rotating motion. Thus the average force distribution what we want exactly is determined by simply multiplying the number of blade with the average force distribution obtained from one blade.

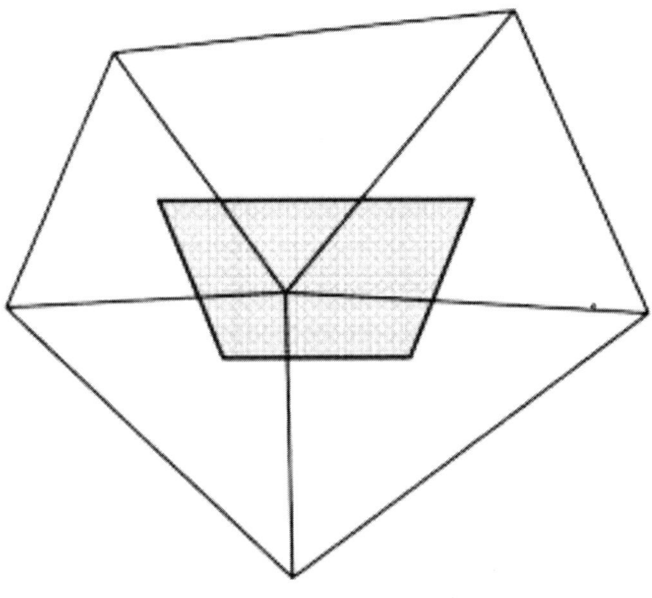

Figure 8. Sketch of an area-weighted interpolation scheme. The shaded trapezoid represents an element on propeller disk and triangles are the projection of surface grid elements of blade on the disk.

Fig. 9 shows computed hydrodynamic force and moment versus J for a range of oblique angles. The force and moment are made dimensionless by

$$T_i' = \frac{T_i}{\rho n^2 D^4}, Q_i' = \frac{Q_i}{\rho n^2 D^5} \left(i = x, y, z\right).$$

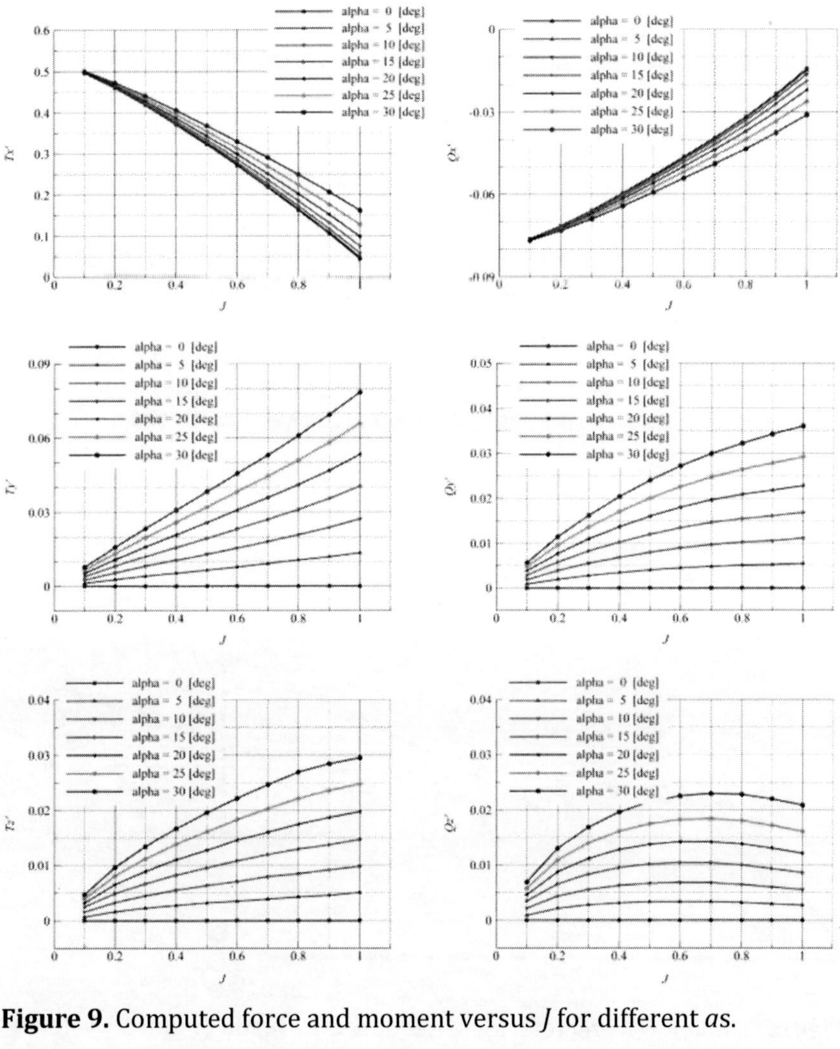

Figure 9. Computed force and moment versus J for different αs.

The propeller efficient is expressed by

$$\eta = \frac{J}{2\pi} \frac{T_x'}{-Q_x'}.$$

As seen in the figure, with the increase of advance ratio the thrust (Tx')decreases but the side force (Ty') and vertical force (Tz') become larger. For the moments, the torque (Qx') decreases as well and (Qy') increases, but (Qz') increases first and then decreases. It is also noted that the influences of oblique angle on force and moment become larger and larger when advance ratio increases, and with the decrease of advance ratio the influences reduce to zero gradually. This is quite understandable and could be explained by making use of velocity vectors for a section of a blade shown in Fig. 10. In the figure nDn represents circumferential velocity. It is no doubt that for a fixed revolution rate the level of the oblique angle impact on loads acting on the blade section is determined by the value of $|a_{kx} - a_{k2}|$, where a_{k1} andak_2 are effective angles of attack. Larger \a$_k$ 1 i\ means

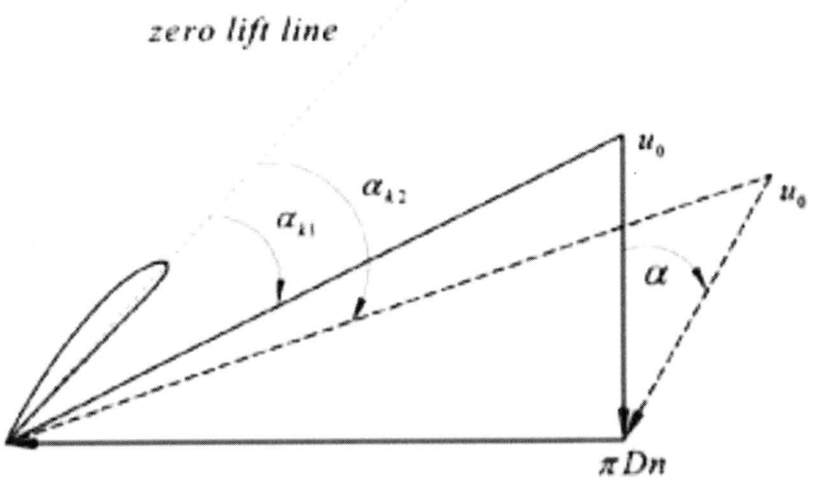

Figure 10. Velocity vectors for a section of a blade (induced velocity neglected).

larger influences, and if $|a_{kx} - a_{k2}|$ is zero no influence occurs. On the other hand, in present considerations large advance ratio corresponds to small revolution rate, since the free-stream velocity u_0 keeps constant and $J = u_0/(nD)$. If oblique angle is fixed and revolution rate goes to infinity, both advance ratio and $|a_{kx} - a_{k2}1$ tend to zero, which means influences of oblique angle on thrust and torque become smaller when advance ratio decreases. In contrast, if revolution rate tends to zero, advance ratio tends to infinity and $|a_{kx} - a_{k2}1$ increases, resulting in larger influences when advance ratio increases. From another point of view, if advance ratio is changed through the free-stream velocity but not the propeller revolution rate, advance ratio could be zero corresponding to zero velocity of free stream. There is of course no influence of oblique angle on thrust and torque when advance ratio is zero.

With above analyses it could be also to explain the patterns of force distributions on propeller disk. Fig. 11 shows the force distributions of the three components of force on the disk for the cases of $J = 0.7$ and $a = 0°, 10°, 20°, 30°$. It is observed that when oblique angle is zero, the axial force distribution is axisymmetric, while for the other two components there exists a pair of region on the disk, in which the distribution pattern is identical but with opposite sign. When oblique angle is not zero, the force distributions are no longer symmetric, and this trend is more obvious as oblique angle increases. It is interesting to find that for the case of $J = 0.7$, $a = 20°$, it appears a region of negative thrust in axial direction and this trend is becoming more and more obvious with the increase of *a*. The appearance of negative thrust is due to that the blade section experiences negative effective angle of attack.

Fig. 12 shows contour plots of pressure distribution on pressure side (left) and suction side (right) for two cases of $J = 0.7$, $a = 0°$ (up) and $J = 0.7$, $a = 30°$ (down). The pressure is non-dimensionalized by $(p - p_0)(0.5\rho u_0^2)$ where $p_0 = 0$ is reference pressure. The pressure distributions on blades could give us a better understanding of the average force distributions on the disk. For the case of $J = 0.7$, $a = 30°$, the blade located at the position of phase angle 150° defined in *o-yz* plane experiences largest loads than other blades on pressure and suction sides, which indicates the blade experiences the largest effective angle of attack. This is accordance with the average thrust distribution of this case shown in Fig. 11.

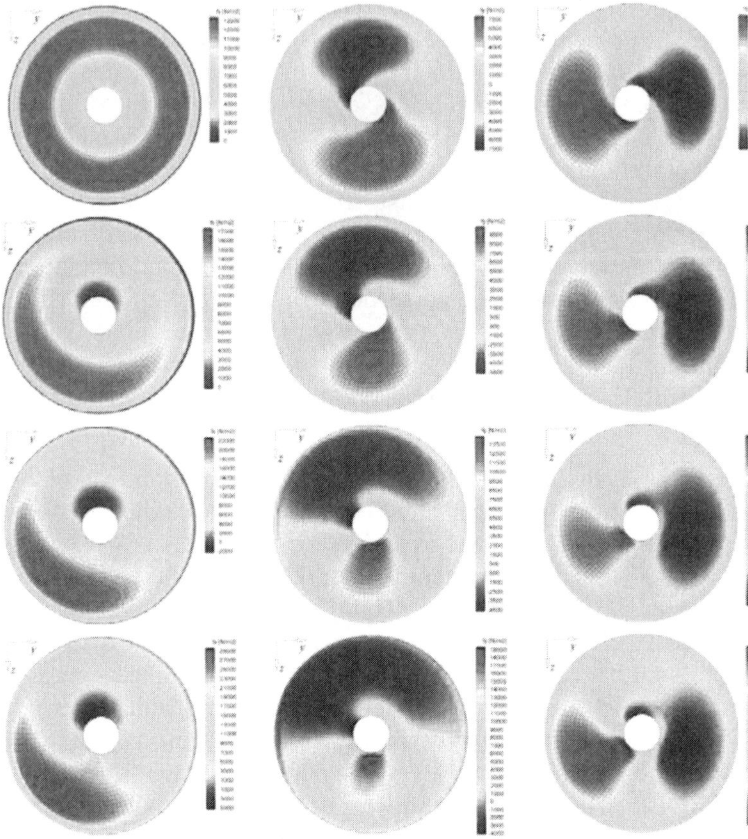

Figure 11. Distributions of the three components of force on propeller disk at $J = 0.7$ and $a = 0°, 10°, 20°, 30°$ (from top to down).

Fig. 13 shows a comparison of iso-surfaces of $Q = 5000\ s^{'2}$ between the cases of $J = 0.7, a = 0°$ (up) and $J = 0.7, a = 30°$ (down). Q is a scalar to visualize turbulent fields. It is defined by

$$Q = -\frac{1}{2}\left(S_{ij}S_{ij} - \Omega_{ij}\Omega_{ij}\right)$$

where

$$S_{ij} = \frac{1}{2}\left(\frac{U_i}{x_j} + \frac{U_j}{x_i}\right), \quad \Omega_{ij} = \frac{1}{2}\left(\frac{U_i}{x_j} - \frac{U_j}{x_i}\right).$$

Figure 12. Pressure distributions on pressure side (left) and suction side (right) at $J = 0.7$, $a = 0°$ (up) and $J = 0.7$, $a = 30°$ (down).

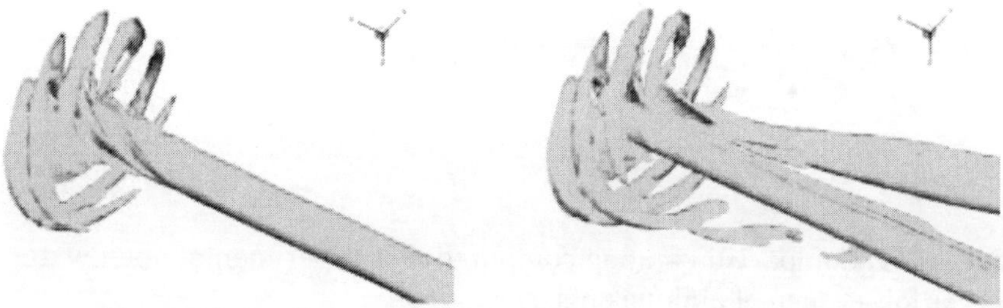

(a) (b)

Figure 13. Comparison of iso-surfaces of $Q = 5000\ s^{-2}$ between the cases of $J = 0.7$, a=0° (a) and $J = 0.7$, $a = 30°$ (b).

As seen, the flow around the propeller becomes not axisymmetric in oblique flow.

Current simulations are validated through comparisons between computed results and experimental data. The computed and experimental open water curves are plotted inFig. 14 for a range of

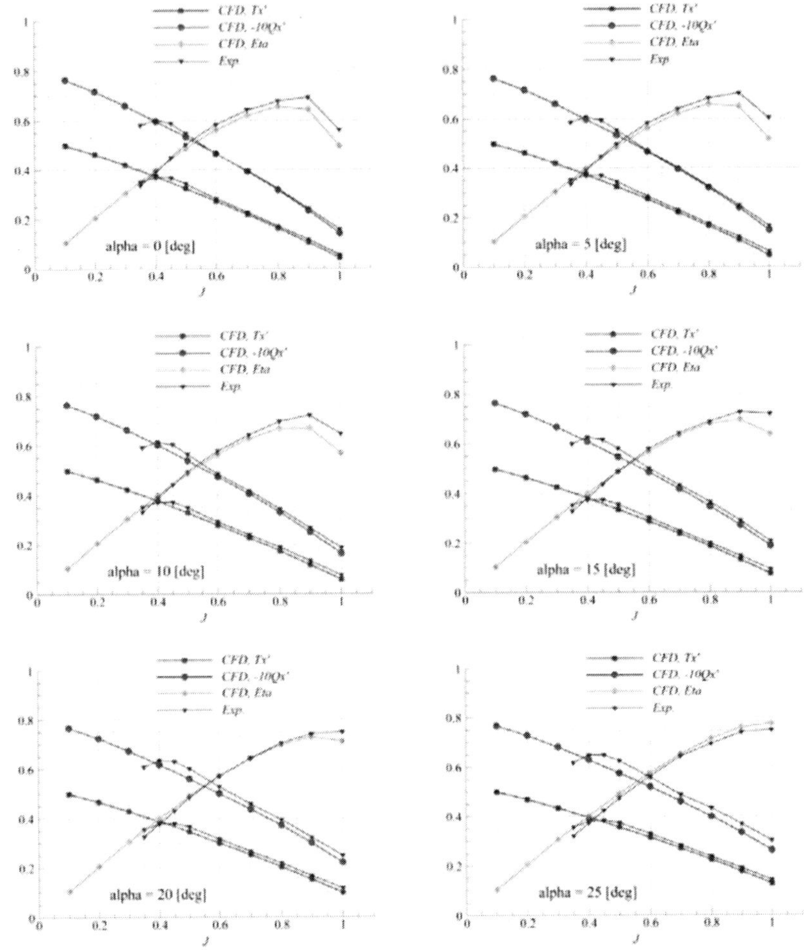

Figure 14. Comparison among computed and experimental open water curves for a range of oblique angles.

oblique angles. It is noted that when advance ratio is larger than 0.45 the RANS results show a generally good agreement with experimental data. However, when advance ratio is smaller than 0.45 the experimental T' and Q'_x become smaller as decrease of advance ratio, while the computed T' and Q'_x show opposite characteristic. This is because when advance ratio is smaller than 0.45 (corresponding to n = 44.44 *rps)* cavitation becomes severe observed in experiments, which causes a thrust loss strongly. However in present study the cavitation is not considered. Although at J = 0.6 and a = 25° a weak cavitation occurs at tips located at upstream as shown inFig. 15, it seems the weak cavitation has no significant impacts on thrust and torque acting on blades since the computed results still show good agreements with experimental data. The comparisons also show that at small oblique angles the RANS results agree excellently with experimental data under no or weak cavitation condition, while at large oblique angles the discrepancy becomes larger. The largest error is found to be around 15%. The large discrepancy at large oblique angle may be caused by the effects of tunnel

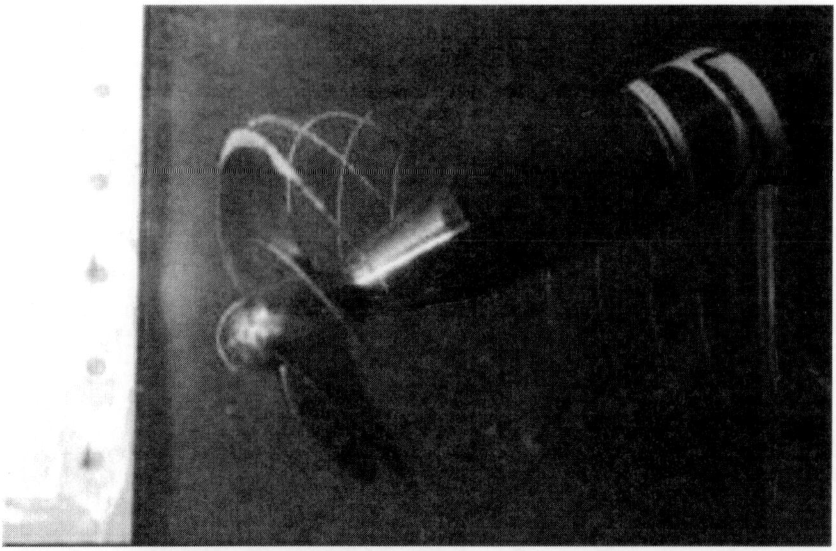

Figure 15. Cavitation observed in experiment at J=0.6 and a=25°.

wall, because the section size of the tunnel is 0.7 *m* x 0.7 m, and at *a* = 25° the distance between propeller center and the closest wall is around 2.5D, which may be not far enough anymore.

Fig. 16 shows the computed and experimental T', T', T' and Q' versus oblique angle for *J*= 0.5 to 1.0. It is seen that T_y and T_z are relatively quite small with respect to T_z. There are obvious differences in T_y and T_z between simulations and experiments, especially for T_z. This is probably due to cavitation.

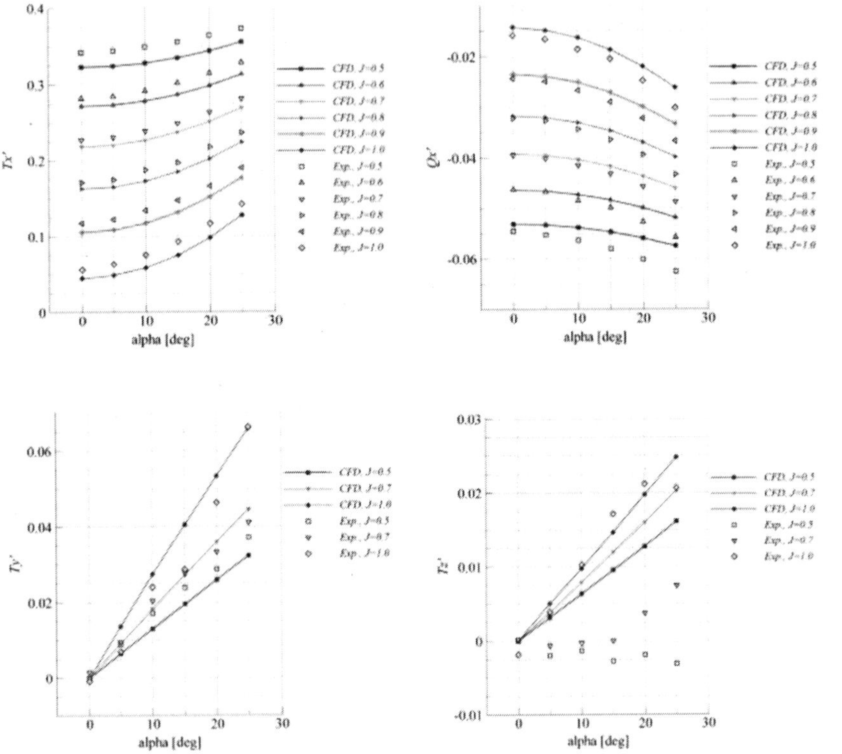

Figure 16. Computed and experimental T', T', T' and Q' versus oblique angle for selected advance ratios.

CONCLUSIONS

This study has used the unsteady RANS solver in OpenFOAM to investigate hydrodynamic performance of a propeller in oblique flow. A sliding grid approach is applied to compute the rotating motion. 70 combinations of advance ratios and oblique angles are considered. The current simulations are validated via comparisons between computed hydrodynamic force, moment and available experimental data. The computed results show generally good agreement with experimental data under no or weak cavitation condition. The cavitation could be considered in future work, which would be more interesting. Generally the present results could be improved by using high revolution grids and large number of time steps in one propeller revolution, for example more than 400. In addition, the averaged force distributions on propeller disk in one revolution are obtained by using an area-weighted interpolation scheme, which is a great foundation to develop better body force models and could give us a better understanding of hydrodynamic performance of the propeller in oblique flow.

ACKNOWLEDGMENTS

This work is financially supported by China Scholarship Council (CSC). The author would like to thank Professor Dr.-Ing. Andrés Cura Hochbaum and Dipl.-Ing. Johannes Beutel for helpful discussions and offering experimental data.

REFERENCES

1. Cura-Hochbaum, A., 2006. Virtual PMM tests for manoeuvring prediction. In Proceeding of the 26th Symposium on Naval Hydrodynamics. Rome, Italy, 17 - 22 September 2006.
2. Dubbioso, G., Muscari, R. and Mascio, A.D., 2013. Analysis of the performances of a marine propeller operating in oblique flow. Computers & Fluids, 75, pp.86-102.
3. Dubbioso, G., Muscari, R. and Mascio, A.D., 2014. Analysis of a marine propeller operating in oblique flow. Part 2: very high incidence angles. Computers & Fluids, 92, pp.56-81.
4. Krasilnikov, V., Zhang, Z.R. and Hong, F.W., 2009. Analysis of unsteady propeller blade forces by RANS. In Proceeding of the First International Sysmposium on Marine Propulsors. Trondheim, Norway, 22-24 June 2009.
5. Menter, F.R., Kuntz, M. and Langtry, R., 2003. Ten years of industrial experience with the SST turbulence model. In Proceeding of Turbulence, Heat and Mass Transfer 4. Antalya, Turkey, 12-17 October 2003.
6. OpenCFD Ltd., 2014. OpenFOAM user guide of version 2.3.0. Bracknell, United Kingdom: OpenCFD Ltd.
7. Shamsi, R. and Ghassemi, H., 2013. Numerical investigation of yaw angle effects on propulsive characteristics of podded

propulsors. International Journal of Naval Architecture and Ocean Engineering, 5, pp.287-301.

8. Simonsen, C.D., Otzen, J.F., Klimt, C., Larsen, N.L. and Stern, F., 2012. Manoeuvring predictions in the early design phase using CFD generated PMM data. In Proceeding of the 29th Symposium on Naval Hydrodynamics. Gothenburg, Sweden, 26-31 August 2012.

9. Stern, F., Agdrup, K., Kim, S.Y., Hochbaum, A.C., Rhee, K.P., Quadvlieg, F., Perdon, P., Hino, T., Broglia, R. and Gorski, J., 2008. Experience from SIMMAN 2008—The first workshop on verification and validation of ship manoeuvring simulation methods. Journal of Ship Research, 55(2), pp.135-147.

9

CHAPTER

RANS Simulation of Cavitation and Hull Pressure Fluctuation for Marine Propeller Operating Behind-Hull Condition

Kwang-Jun Paik, Hyung-Gil Park and Jongsoo Seo

Samsung Ship Model Basin (SSMB), Samsung Heavy Industries Co., Ltd., Daejeon, Korea

ABSTRACT

Simulations of cavitation flow and hull pressure fluctuation for a marine propeller operating behind a hull using the unsteady Reynolds-Averaged Navier-Stokes equations (RANS) are presented. A full hull body submerged under the free surface is modeled in the computational domain to simulate directly the wake field of the ship at the propeller plane. Simulations are performed in design and ballast draught conditions to study the effect of cavitation number. And two propellers with slightly different geometry are simulated to validate the detectability of the numerical simulation. All simulations are performed using a commercial CFD software FLUENT. Cavitation patterns of the simulations show good agreement with the experimental results carried out in Samsung CAvitation Tunnel (SCAT). The simulation results for the hull pressure fluctuation induced by a propeller are also compared with the experimental results showing good agreement in the tendency and amplitude, especially, for the first blade frequency.

KEYWORDS: Propeller; Cavitation; Hull pressure fluctuation; RANS; Propeller-hull interaction

INTRODUCTION

The practical importance of accurate numerical simulations to predict the cavitation pattern and hull pressure amplitude has increased in the field of ship building industries. The researches to simulate the cavitation flow using RANS solver were performed actively in the last several years because cavitation models for RANS solver and computation power have been rapidly developed in the last decade.

Watanabe et al. (2003) performed a pioneering study on the cavitation flow for a marine propeller using RANS. Salvatore et al. (2009) compared seven computational models using RANS, LES, and BEM for the INSEAN E779A propeller in the uniform flow and the wake field. The cavitation simulation for conventional and highly-skewed propellers in the behind-hull condition using an in-house RANS solver was performed by Shin et al. (2011). Hasuike et al. (2010) simulated the cavitation flow and showed a possibility to predict the cavitation erosion risk with the indexes using time differential of cavity void fraction or pressure.

On the other hand, the usage of commercial CFD software gradually increased to simulate the cavitation flow for marine propellers. Boorsma and Whitworth (2011) used STAR-CCM+ to study on the prediction of cavitation and erosion for a propeller and rudder. Bertetta et al. (2011) compared the results of RANS solver using STAR-CCM+ with those of potential solver for the cavitation flow of a controllable pitch propeller.Morgut and Nobile (2011) studied on the performance of three cavitation models for a marine propeller using ANSYS-CFX. Li and Terwisga (2011) used FLUENT to investigate unsteady cavitation phenomena for 2D and 3D foils. Kawamura (2010) assessed cavitation erosion risk based on pressure impacts and simulated the cavitation flow and the amplitude of pressure fluctuation for a marine propeller using FLUENT.

Most of simulations except for Kawamura (2010) prescribed a velocity distribution at the inlet boundary of the computational domain to achieve the wake field at the propeller plane. In that case, the adjustment of inlet boundary condition to get a desired velocity contour at propeller plane is of primary importance and very laborious work. Therefore,Kawamura (2010) used a full hull body to generate the wake field at propeller plane. Nevertheless, the cavity extent was under-predicted and the pressure amplitudes were about 70% of the experimental results.

In this paper the results of numerical simulations using RANS for the cavitation pattern and hull pressure fluctuation of marine propellers are presented. A commercial CFD software FLUENT version 14.0 is used for the numerical simulations. The cavitation model of Schneer and Sauer (2001) is applied to simulate the cavitation flow. A computational domain including a full hull body submerged under the free surface is used to simulate directly the wake field at the propeller plane. Two kinds of studies varying cavitation number and propeller geometry are executed in order to validate the numerical simulation method. And the numerical simulations are performed at the same as the test conditions in the cavitation tunnel and compared with the experimental results.

NUMERICAL METHODS AND MODELS

A commercial CFD software FLUENT version 14.0 was used for the numerical simulations, in which the cavitation flow was solved by a mixture model based on a single-fluid multiphase mixture flow approach.

Governing equations

In the mixture model, the continuity equation and the momentum equation become as

$$\frac{\partial}{\partial t}(\rho_m) + \nabla \cdot (\rho_m \vec{v}_m) = 0 \tag{1}$$

$$\frac{\partial}{\partial t}(\rho_m \vec{v}_m) + \nabla \cdot (\rho_m \vec{v}_m \vec{v}_m) = -\nabla p + \nabla \cdot \left[\mu_m \left(\nabla \vec{v}_m + \nabla \vec{v}_m^T\right)\right] + \rho_m \vec{g} + \vec{F} \tag{2}$$

The mixture density and viscosity coefficient are defined as

$$\begin{aligned} \rho_m &= \alpha \rho_v + (1-\alpha)\rho_l \\ \mu_m &= \alpha \mu_v + (1-\alpha)\mu_l \end{aligned} \tag{3}$$

where α is the vapor volume fraction. Subscripts m, v, and l represent mixture, vapor, and liquid phase, respectively.

Cavitation model

The cavitation model of Schneer and Sauer (2001) applied in this research solves the vapor volume fraction with following transport equation:

$$\frac{\partial}{\partial t}(\alpha \rho_v) + \nabla \cdot (\alpha \rho_v \vec{v}_v) = R_e - R_c \tag{4}$$

where the terms in the right hand side, R_e and R_c, account for the mass transfer between the liquid and vapor phase in cavitation denoting the evaporation and condensation of the vapor bubbles. The forms of R_e and R_c are written as follows:

when $p_v \geq p$,

$$R_e = \frac{\rho_v \rho_l}{\rho_m} \alpha(1-\alpha) \frac{3}{R_B} \sqrt{\frac{2}{3} \frac{(p_v - p)}{\rho_l}} \tag{5}$$

when $p_v \leq p$,

$$R_c = \frac{\rho_v \rho_l}{\rho_m} \alpha(1-\alpha) \frac{3}{R_B} \sqrt{\frac{2}{3} \frac{(p - p_v)}{\rho_l}} \tag{6}$$

Here, the bubble radius, R_B, is expressed as

$$R_B = \left(\frac{\alpha}{1-\alpha} \frac{3}{4\pi} \frac{1}{n_b} \right)^{\frac{1}{3}} \tag{7}$$

where n_b is the bubble number density. The source terms, R_e and R_c, in the transport equation approach zero when $\alpha = 0$ and $\alpha = 1$.

MODEL TEST

Model tests to observe cavitation flow on a propeller blade and to measure pressure fluctuation on a hull surface induced by the propeller cavitation were carried out in Samsung CAvitation Tunnel (SCAT). The principal particulars of the test section of SCAT are summarized in Table 1 and the set-up of a model ship and propeller is shown in Fig. 1. The definition of propeller blade angle is depicted in Fig. 2. The propeller blade angle begins from the top position toward propeller rotation direction. Pressure transducers to measure the pressure fluctuation on a hull surface were installed on the top of propeller with the intervals of 30 *mm* as illustrated in Fig. 3.

Table 1. Principal particulars of the test section of SCAT.

Item	Value
Dimension of test section (L × B × D)	12.0 × 3.0 × 1.4 *m*
Maximum speed	12.0 *m/s*
Contraction ratio	2.75

Figure 1. Set-up of model ship and propeller in SCAT.

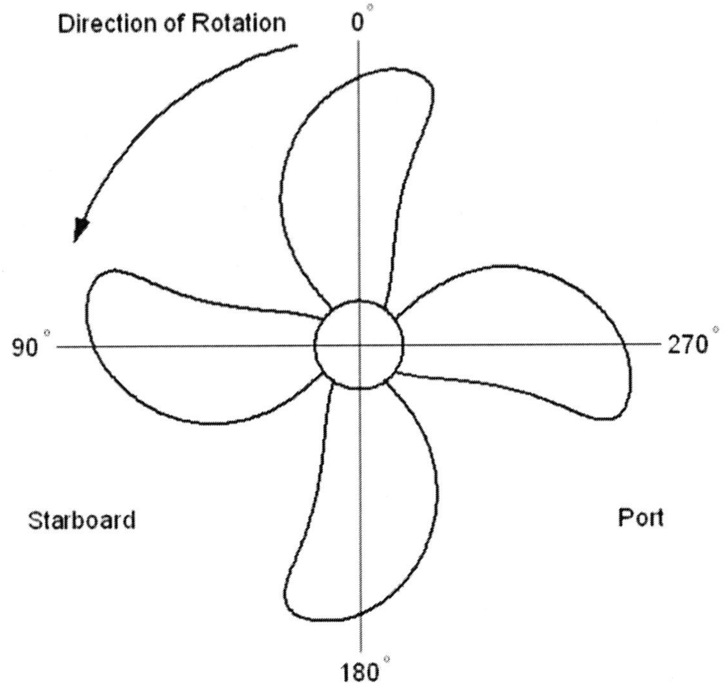

Figure 2. Definition of propeller blade angle.

The length of the model ship, a crude oil tanker, was about 7.6 *m* and the diameter of the model propeller was 226.1 *mm* in this research. The vertical distance between propeller tip and the pressure transducer P3 is about 30% of propeller diameter. The blockage of the test section due to the model ship was about 13.2%. The surface of the model propeller was coated with a rough paint to stabilize the behavior of cavitation flow. The inflow velocity of model test was fixed as 5.5 *m/s*. The propeller rotation speed was set with the thrust identity method based on the results of self-propulsion test carried out in a towing tank.

NUMERICAL SIMULATION

Grid system

To simulate the propeller cavitation flow under similar situation to the model test, a full hull body submerged under the free surface was modeled in the computational domain. The free surface was substituted with a symmetry boundary condition because the mixture model for the cavitation flow and the Volume of Fluid (VOF) model for the free surface could not be implemented simultaneously in FLUENT.

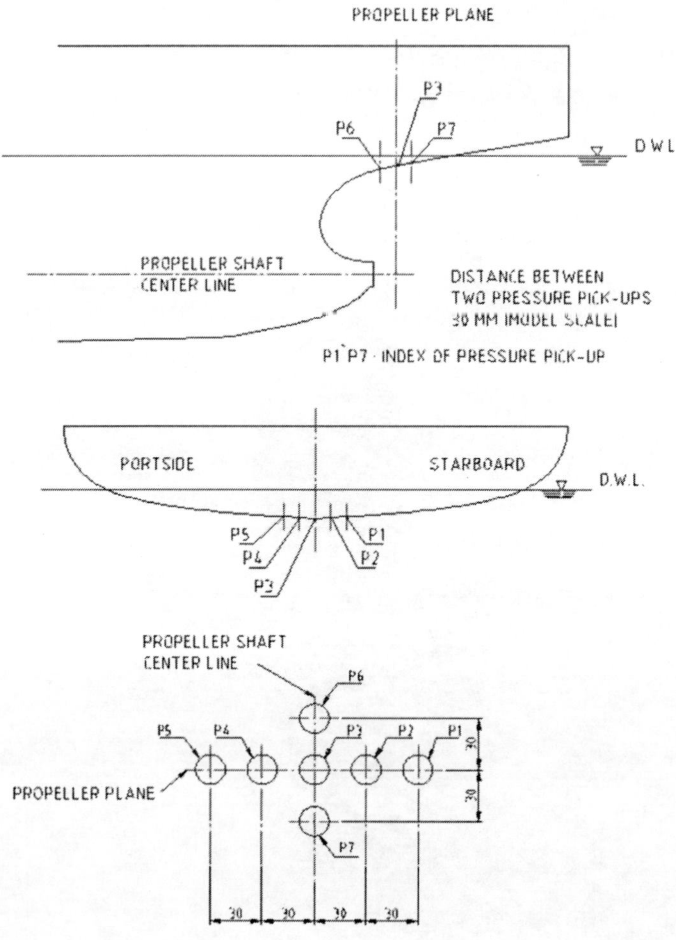

Figure 3. Position of pressure transducers installed on a model ship.

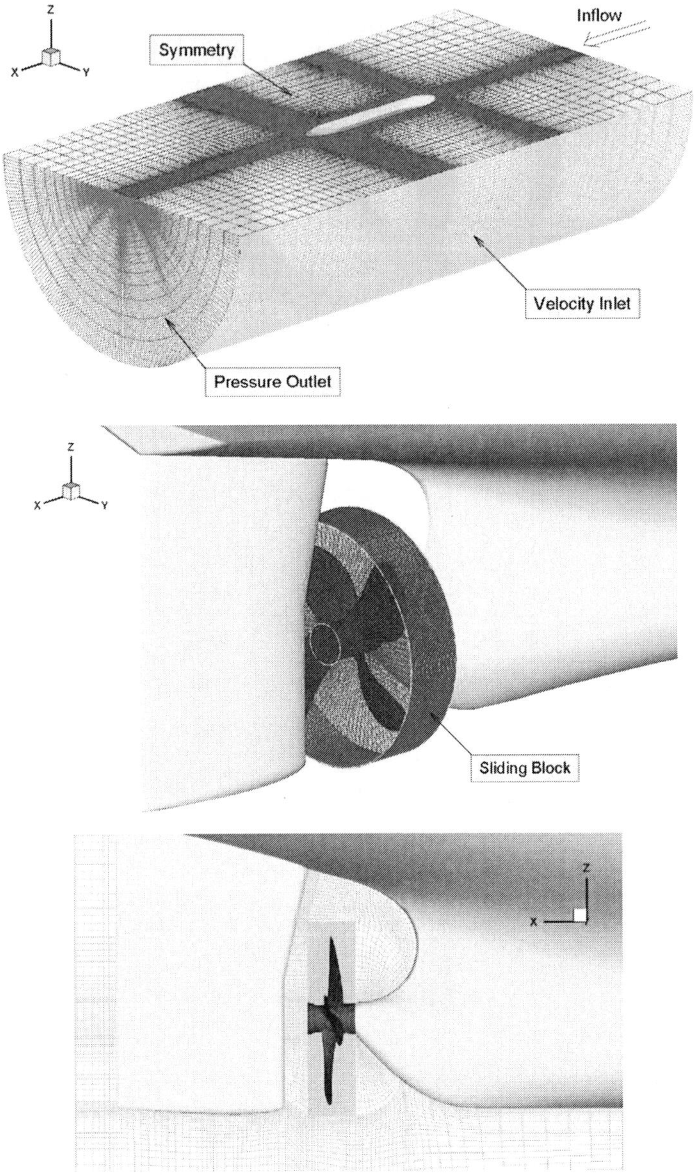

Figure 4. Outline of grid system for computational domain (top left) and propeller block (top right) and x-z plane section showing the interface between sliding and outer blocks (bottom).

A sliding block surrounding the propeller, composed of unstructured girds, was applied to implement the effect of propeller rotation. Pyramid cells were used for the surface of propeller and boundaries, and tetrahedral cells were filled in the block. On the other hand, structured girds were applied to the other domains expect for the sliding block. Fig. 4 shows the outline of grid system including boundary conditions and the interface between sliding and other blocks. While the test section of the cavitation tunnel was rectangular, a half circular sectional profile was used for the computational domain to achieve a good grid quality of structured grid. The grid sizes were 1.4 *M* and 2.6 *M* cells for the sliding block and the other blocks, respectively.

Simulation method

As the numerical model for cavitation flow, Schneer and Sauer (2001) was applied in this research. Bubble number density was set as 1e+15, which corresponds to the bubble radius of 3 μm and 13 μm at α = 0.1 and α = 0.9, respectively. To improve the convergency of cavitation flow and reduce the computational time, Multiple Reference Frame (MFR) was used at first and Sliding Mesh Model (SMM) was then applied to simulate the rotation of propeller. Propeller rotated with a constant time step corresponding to the rotating angle of 1.5° to obtain 60 data per propeller blade phase. Time histories of fluctuating pressure on the hull surface were recorded to compare with the experimental results at the same positions as the pressure transducers installed on the model test. The simulation methods applied in this research are summarized in Table 2. The inlet boundary condition was set as constant velocity with 3% and 200 for turbulent intensity and turbulent viscosity ratio, respectively. The wall y+ on hull and propeller were about 130 and 200.

Table 2. Summary of simulation methods.

Item	Method
Governing equation	RANS
Turbulence model	RSM w/ standard wall function
Cavitation model	Mixture
P-V coupling	SIMPLEC
Pressure solver	Standard
Momentum solver	Second order upwind

A propeller coated with the rough paint tended to increase the cavity extent and pressure fluctuation as compared with a smooth surface propeller in the experiment. However, since it was impossible that the effect of the paint was implemented in numerical simulation, a smaller cavitation number was used for the simulation instead of the rough paint. The adjustment was approached with ignoring a wave height at stern which was considered in the model test.

RESULTS AND DISCUSSION

In this research two kinds of simulations were executed and compared with the experimental results. The first simulation was performed at two draught conditions to investigate the effect of cavitation number. The other study was to validate the detectability of the numerical simulation about the differences of cavitation pattern and pressure amplitude from the small modification of propeller geometry.

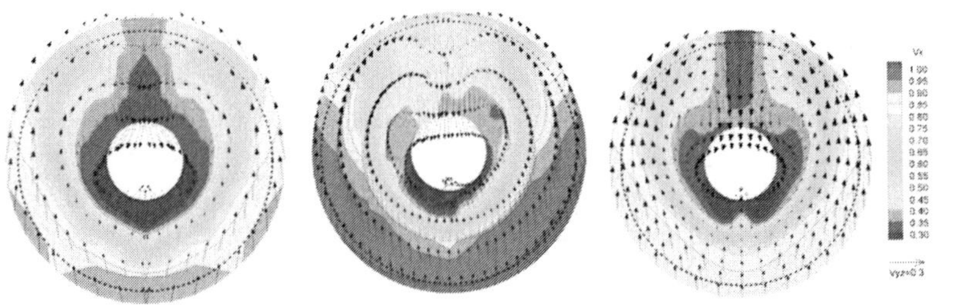

Figure 5. Velocity contours and vectors of experiments (left: towing tank, middle: cavitation tunnel) and simulation (right) at propeller plane (dashed circle: outline of propeller disk).

Prior to the cavitation simulation, the bare hull without the propeller was simulated to compare the wake field at the propeller plane. Fig. 5 shows the comparison of velocity contours and vectors between the experimental data and simulation result at the propeller plane. Model ship speed was 1.35 *m/s* in the towing tank test. Inlet velocity of the numerical simulation was 5.5 *m/s*, which was the same velocity as the cavitation tunnel. The velocity contour and vector are normalized by the ship speed or inlet velocity. The wake velocities and contours at the towing tank and simulation are very similar, while the axial velocity at the cavitation tunnel is about 0.1 faster at outer radii of propeller disk. It means that the velocity at propeller plane in the cavitation tunnel is probably contracted and accelerated due to the blockage effect. The variation of axial velocity at the cavitation tunnel is not smooth around 20° and 340°.

Draught condition

The simulations were performed at design and ballast draught conditions, and the simulation results were compared with the experimental results of SCAT. The propeller tested in the model test is a 4-blade propeller (Prop-A) designed for a crude oil tanker. Mean pitch ratio and expanded area ratio of the propeller are 0.655 and 0.480, respectively. The simulation conditions are summarized in Table 3. The cavitation number is defined as

$$\sigma = \frac{p - p_v}{0.5\rho n^2 D^2} \quad (8)$$

where p is a static pressure at the 70% of propeller radius above the propeller shaft center and p_v is a vapor pressure. A propeller rotation speed and a propeller diameter are denoted as n and D, respectively.

Table 3. Summary of simulation conditions to study the effect of draught condition.

	Design	Ballast
Inflow speed	5.5 *m/s*	
Propeller diameter	226.1 *mm*	
Propeller speed	38.0 *rps*	39.3 *rps*
Cavitation number	2.129	1.462

The thrusts of the simulations at the design and ballast conditions were about 4% higher than those of the experiments. The larger thrusts of the simulations result from the slower axial velocity at the propeller plane as compared in Fig. 5.

Cavitation patterns at the propeller blade angle of 0°, 20°, and 40° at the design and ballast draught conditions are compared with the experimental results in Figs. 6 and 7. The cavity extent in the simulation are expressed with the isosurface of $\alpha = 0.1$. The simulation results have some limitations to capture the tip vortex cavitation, but the cavity patterns generally well agree with the experiments. At the design draught condition, the leading edge cavitation near 0.8 *R* of the propeller blade is not appeared in the numerical simulation, while it is observed with very unstable behavior in the experiment. At the ballast draught condition, the cavity extents at all blade angles well correspond with the experiments. The sheet cavitation was not appeared in the range between 90° and 330° in the experiments, but it was observed until about 240° in the simulations due to the slower axial velocity than the experiment as shown in Fig. 5.

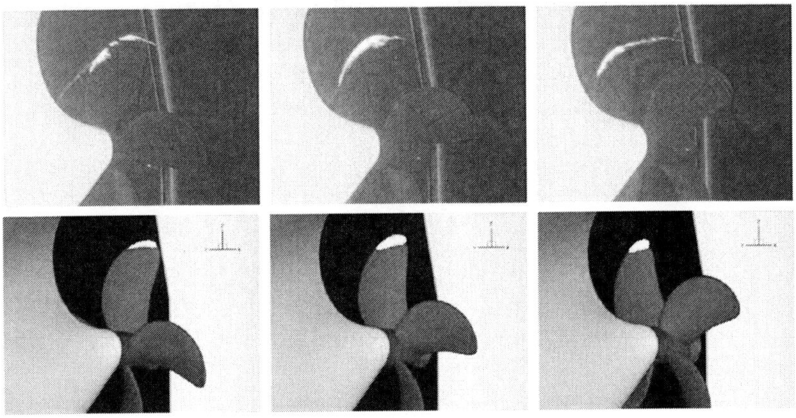

Figure 6. Comparison of cavitation patterns at design draught condition for Prop-A (top: EFD, bottom: CFD; left: 0°, middle: 20°, right: 40°)

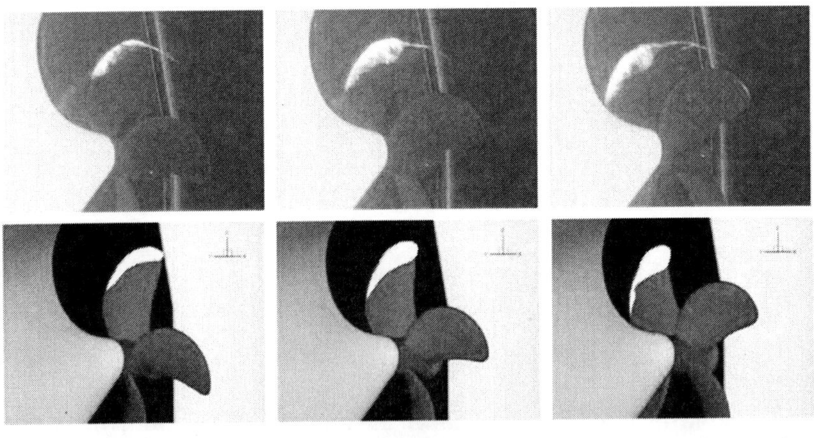

Figure 7. Comparison of cavitation patterns at ballast draught condition for Prop-A (top: EFD, bottom: CFD; left: 0°, middle: 20°, right: 40°).

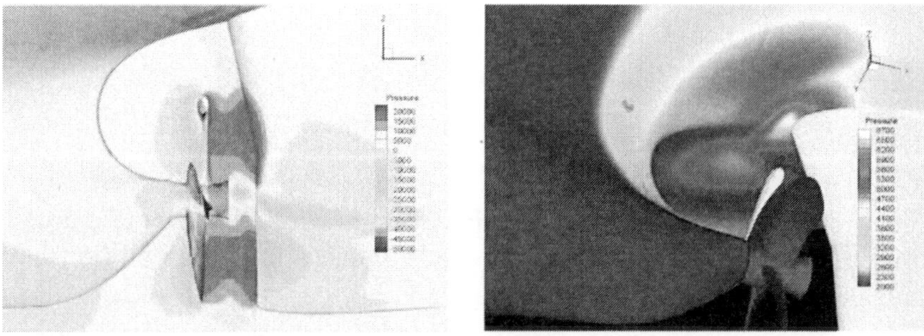

Fig. 8. Instantaneous pressure distributions in x-z plane section including hull and propeller surfaces (left) and hull surface pressure above propeller position (right) during propeller operation.

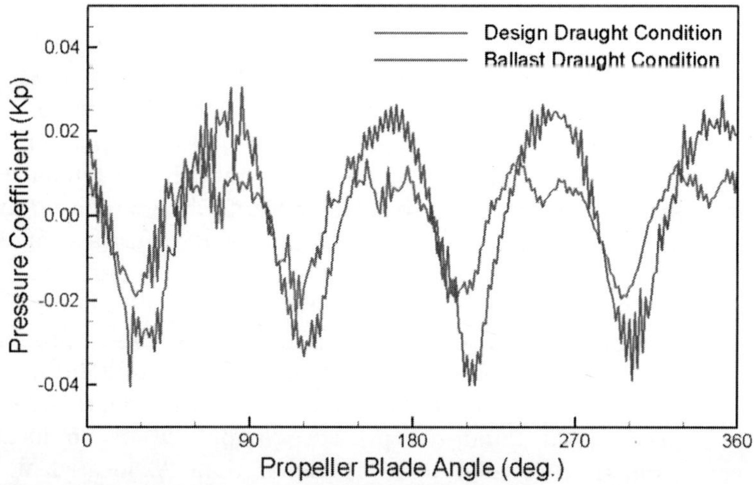

Figure 9. Comparison of pressure fluctuations of P2 transducer at ballast and ballast draught conditions for Prop-A.

Fig. 8 shows an instantaneous pressure distribution on the hull surface induced by the propeller cavitation as well as field velocity around propeller in x-z plane section. Very high pressure concentrates above the propeller position, especially on starboard side rather than port side. Time histories of pressure fluctuation on P2 transducer at the design and ballast draught conditions are plotted in Fig. 9. Hull pressure is expressed with the pressure coefficient defined as

$$K_P = \frac{\Delta p}{\rho n^2 D^2} \tag{9}$$

where Δp is the pressure fluctuation amplitude from the mean pressure. The pressure fluctuation at the design draught condition has eight peaks during one revolution, while that at the ballast draught condition has four peaks.

The hull pressure amplitudes for Prop-A at the design draught condition are compared with the experimental results in Fig. 10. Since the position of P7 was concealed beneath the rudder trunk, P7 could not be measured in this research. For the first blade frequency (1BF), the simulation results show good agreement with the experimental results in the tendency and magnitude of pressure fluctuation, but the pressure amplitudes of P1 and P2 at starboard side are about 20% and 10% higher than the experiments, respectively. At the other pressure transducers, the pressure amplitudes are very similar to the experiments. The second blade frequencies (2BF) of the simulation are almost half of the experimental results, even though the tendency is similar to the experiments. As explained in Fig. 5, the gradient of axial velocity around 20° and 340° probably affects the increase of 2BF amplitude in the experiment. In contrast, the velocity gradient in the simulation is very smooth, resulting in relatively small amplitude of 2BF. And the numerical diffusion and grid fineness around the propeller tip might be another reason for smaller 2BF amplitude in the simulation. Overall, the hull pressure amplitudes of starboard side are higher than port side, and the pressure amplitude of P6 located upstream from the propeller plane is higher than P3 located at the propeller plane.

Fig. 11 shows the comparison of the hull pressure amplitudes at the ballast draught condition with the experimental results. The amplitudes of 1BF are about 10% higher than the experiments except for P1 and P6. The pressure amplitude of P1 is about 28% higher than the experiment, while that of P6 is about 5% lower. Nevertheless, the tendency of pressure amplitude is very close to the experiment. However, the amplitudes of 2BF are significantly lower than the experiments because the effect of axial velocity gradient in the simulation may be diminished due to the relatively large cavity extent in the ballast draught condition. This phenomenon can be explained through Fig. 9. The two peaks with the interval of about 40° at the crest of the signal observed in the design load condition are not observed in the ballast condition.

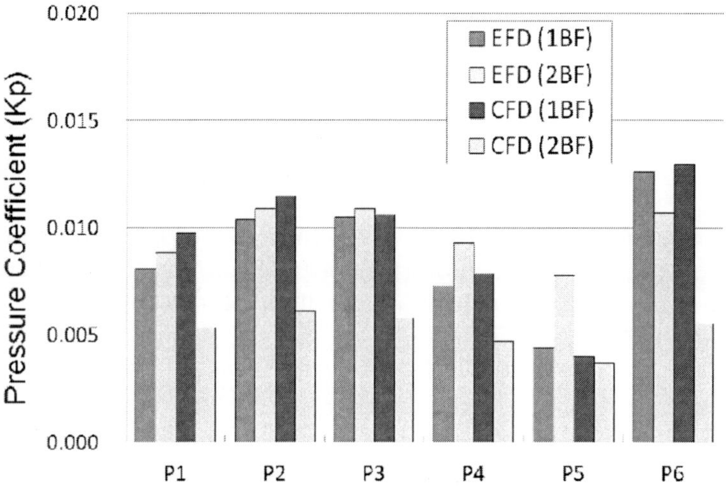

Figure10. Comparison of pressure amplitudes at design draught condition for Prop-A.

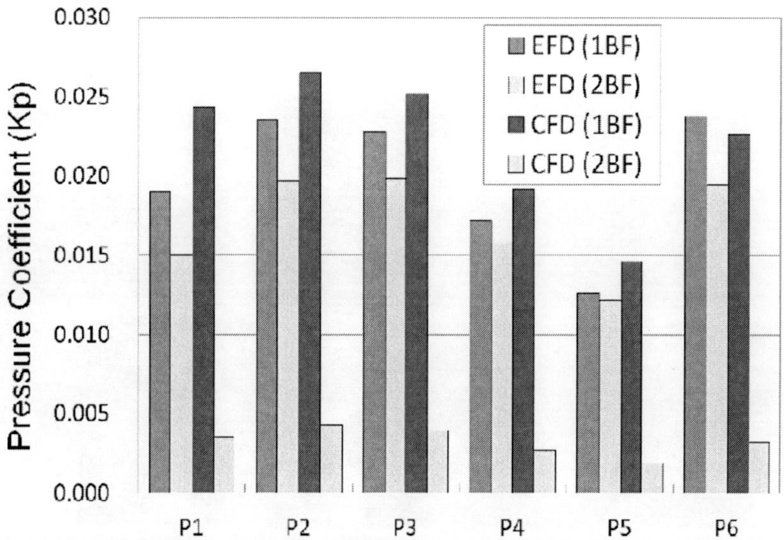

Figure 11. Comparison of pressure amplitudes at ballast draught condition for Prop-A.

Propeller geometry

Another propeller (Prop-B) was designed to study on the effect of propeller rake distribution. Prop-B has the backward rake of 0.014D at tip, while Prop-A has the forward rake of 0.007D at tip. The other parameters for propeller geometry such as pitch, camber, and chord including propeller diameter are exactly same. Due to the rake distribution, the rotation speed of Prop-B is slightly higher than that of Prop-A; as a result, the cavitation number of Prop-B is relatively lower than that of Prop-A. The simulation conditions for Prop-A and Prop-B are summarized in Table 4.

Table 4. Summary of simulation conditions to study the effect of propeller geometry.

	Prop-A	Prop-B
Inflow speed	5.5 *m/s*	
Propeller diameter	226.1 *mm*	
Propeller speed	39.3 *rps*	40.6 *rps*
Cavitation number	1.462	1.455

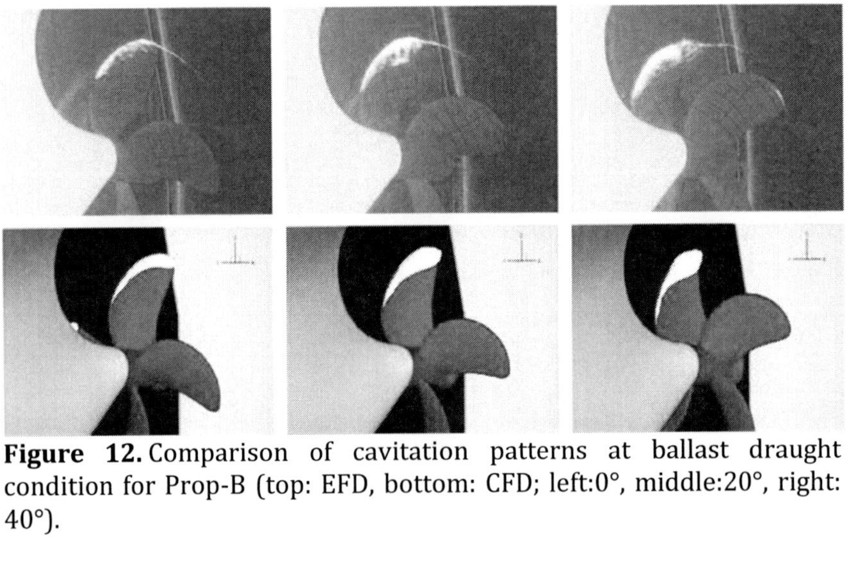

Figure 12. Comparison of cavitation patterns at ballast draught condition for Prop-B (top: EFD, bottom: CFD; left:0°, middle:20°, right: 40°).

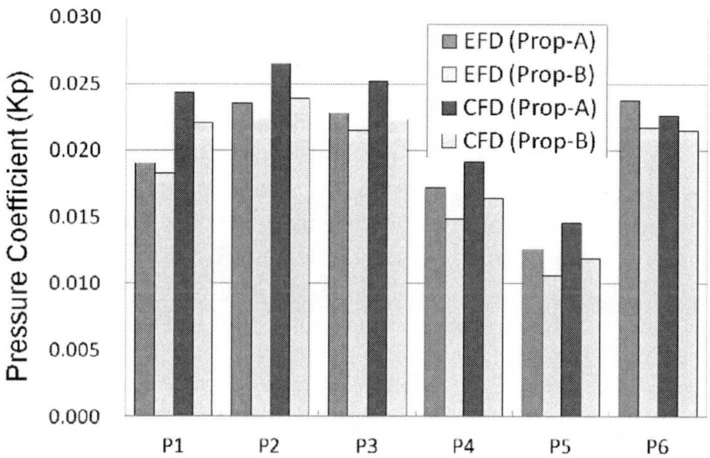

Fig. 13. Comparison of 1BF pressure amplitudes between Prop-A and Prop-B at ballast condition.

The cavitation patterns at the ballast draught condition for Prop-B are compared with the experimental results in Fig. 12. The cavity extents of Prop-B are not different from Prop-A in both experiment and simulation. However, the effect of propeller rake is distinguished in the hull pressure amplitude as shown in Fig. 13. Prop-B with the backward rake reduces the pressure amplitude of about 5% at P1 to P3 and about 13% at P4 and P5 in the experiment. Similarly, the amounts of reduction are about 10% at P1 to P3, about 5% at P4 and about 10% at P5 in the simulation. From the simulation results, it can be concluded that the numerical simulation using RANS is useful to detect the difference of pressure amplitude due to the small modification of propeller geometry.

CONCLUSIONS

Numerical simulation studies using RANS for cavitation flow and hull pressure fluctuation of marine propellers were presented in this paper. A computational domain including a full hull body submerged under the free surface was used to simulate directly the wake field at the propeller plane. And the simulation results were compared with the experimental results performed in the cavitation tunnel.

Two kinds of studies were executed in this research. At first, the numerical simulations were performed in design and ballast draught conditions to study the effect of cavitation number. The cavitation patterns at both conditions agreed well with the experimental results. The first blade frequencies of the pressure fluctuations in the simulations were slightly larger than the experiments, whereas the tendency of the pressure amplitudes according to the location of the transducers was very similar to the experiments. However, for the second blade frequencies the simulations showed some limitations to predict the magnitude of pressure fluctuation due to the difference of wake field at the propeller plane. The other study was to validate the detectability of the numerical simulation for the cavitation pattern and hull pressure amplitude. Two propellers with slightly different geometry were simulated and compared with the experimental results. The cavity extents of the simulation and experiment were not different between two propellers. However, the pressure amplitudes of two propellers had small difference, and that was validated in the numerical simulation.

In conclusion, the cavitation simulation using RANS showed reliable performance to predict the cavitation pattern and hull pressure amplitude through this research. To improve the accuracy of 2BF harmonics in the simulation, however, numerical methods to achieve more accurate wake field at the propeller plane need to be studied in detail. For the further work, various ship types will be simulated and compared with experimental data to evaluate this numerical simulation method.

ACKNOWLEDGEMENTS

This work was partially carried out in the research grant, Development of New Propulsion System for Fuel Saving of Ships (No. 2011-10040081), funded by the Korean Ministry of Knowledge Economy.

REFERENCES

1. Bertetta, D., Brizzolara, S., Gaggero, S., Savio, L. and Viviani, M., 2011. Numerical and experimental characterization of a CP propeller unsteady cavitation at different pitch settings. Second

International Symposium on Marine Propulsors, Hamburg, Germany, 15-17 June 2011.

2. Boorsma, A. and Whitworth, S., 2011. Understanding details of cavitation. Second International Symposium on Marine Propulsors, Hamburg, Germany, 15-17 June 2011.

3. Hasuike, N., Yamasaki, S., Ando, J. and Okazaki, A., 2010. Numerical study on cavitation erosion risk of marine propellers operating in wake flow, International Propulsion Symposium, Okayama, Japan, 19-20 April.

4. Kawamura, T., 2010. Numerical simulation of propulsion and cavitation performance of Marine propeller. International Propulsion Symposium. Okayama, Japan, 19-20 April 2010.

5. Li, Z. and Terwisga, T., 2011. On the capability of multiphase RANS codes to predict cavitation erosion. Second International Symposium on Marine Propulsors, Hamburg, Germany, 15-17 June.

6. Morgut, M. and Nobile, E., 2011. Influence of the mass transfer model on the numerical prediction of the cavitating flow around a marine propeller. Second International Symposium on Marine Propulsors, Hamburg, Germany, 15-17 June 2011.

7. Salvatore, F., Streckwall, H. and van Terwisga, T., 2009. Propeller cavitation modelling by CFD - results from the VIRTUE 2008 Rome workshop. First International Symposium on Marine Propulsors, Trondheim, Norway, 22-24 June.

8. Shin, K. W., Anderson, P. and Mikkelsen, R., 2011. Cavitation simulation on conventional and high-skewed propellers in the behind-hull condition. Second International Symposium on Marine Propulsors, Hamburg, Germany, 15-17 June.

9. Schneer, G. H. and Sauer, J., 2001. Physical and numerical modeling of unsteady cavitation dynamics. 4th International Conference on Multiphase Flow, New Orleans, United States, 27 May - 1 June 2001.

10. Watanabe, T., Kawamura, T., Takekoshi, Y., Maeda, M. and Rhee, S. H., 2003. Simulation of steady and unsteady cavitation on a marine propeller using a RANS CFD code. Fifth International Symposium on Cavitation (CAV2003), Osaka, Japan, 1-4 November 2003.

10
CHAPTER

Numerical Investigation of the Performance of Voith Schneider Propulsion

Ehsan Esmailian[1], Hassan Ghassemi[1], Seyyed Abbas Heidari[1]

[1]*Department of Ocean Engineering, Amirkabir University of Technology, Tehran, Iran*

ABSTRACT

The research explained in this article was carried out to investigate the numerical hydrodynamic characteristics of the Voith Schneider Propeller (VSP) system. The system has four vertical blades connected to the horizontal disk. Each blade has two speeds, one around a disk and second local rational along its axis. The method is finite volume CFD code (Fluent v14) with RNG k-ε turbulence model. For the purposes of this research, a VSP with 4 blades is analyzed at various operating conditions. Hydrodynamic characteristic parameters including moment, thrust and efficiency are presented as functions of advance velocity coefficients at three different pitch rations.

Keywords: Voith Schneider Propeller, CFD, RNG turbulence model, hydrodynamic performance

1. INTRODUCTION

VSP was developed as a high performance drive system by Australian engineer Ernst Schneider and came into operation in 1927. It is used primarily for ships that have to satisfy particularly demanding safety and maneuverability requirements [1]. The thrust is generated by separately oscillating, balanced propeller blades. VSP permits continuously variable thrust adjustment through 360°. Thus, the VSP has no preferential direction of thrust and allows infinite variation in the magnitude and direction of thrust [2, 3].

A significant difference between the VSP and the screw propeller is the direction of the axis of rotation relative to the direction of thrust. In the case of screw propeller, the axis of rotation and the direction of thrust are identical, on the VSP they are perpendicular to one another. The VSPs operate at a comparatively low rotational speed and are therefore notable for their long service life and very low maintenance [5]. Since the VSP simultaneously generates propulsion and steering forces, there is no need for additional appendages such as propeller brackets, rudders, pods, shafts, etc. Blades of VSP are symmetric. In certain installation conditions, the rectangular swept area of a VSP is approximately twice as large as that of a screw propeller. Thus, generated thrust force in the VSP is more than a screw propeller. The perpendiculars of the chords of the profiles intersect at a single point, the steering center N, as the blades revolve. The pitch of VSP can be varied within a range e≤0.8. Propellers with a pitch e>1 are known as trochoidal propellers. The Kirsten-Boeing propeller is a special casewith a pitch of e=1 [5, 7].

Rotation speed in VSP is one fourth of typical propellers. Their diameter varies between 1.2 and 3.8 meters and the input power to them is up to 4100 kW. Parameters affecting hydrodynamic performance of VSP are angle of attack during rotation of the rotor, geometry of blades profile, ratio of blade chord length to diameter of rotor, position of blade shafts, relative thickness of blades, tail-end flap of blades and ratio of blade length (span) todiameter of rotor. The tail-end flap in blades can lower the induced drag force [8, 9, 10].

Assessment of cavitation behavior shows that the sheet cavitation is more common in these propellers. Sheet cavitation has no erosion and corrosion on the blades. VSPs are mainly used in Offshore Support Vessels (OSV), Voith Water Tractor (VWT), Tugs, Mine Countermeasures Vessels (MCMV) and Double-Ended Ferries (DEF) [11, 12].

Design improvement is an important phase of the engineering design process. In reference[4],the numerical optimization and Computational Fluid Dynamics (CFD), in particular the Vortex-Lattice Method for the optimization of VSP is used. In reference[21], the Automatic Differentiation method is used for the optimization of the VoithSchneiderPropeller. In reference[22], a new vertical axis propulsion system with orbital paddles which is constituted by a pair of contra-rotating impellers is proposed.Here, the effect of pitch ratio on efficiency and other hydrodynamic characteristics is investigated. It will be shown that the efficiency increases with pitch ratio.

2. COMPONENT OF ROTOR IN VSP

The energy required to generate the thrust is supplied to the rotor casing (1) via the flanged-on reduction gear (7) and the bevel gear (6) (Figure 1). Gland bearings or special roller bearing are used to support the blade shaft. The rotor casing is axially supported by the thrust plate (10) and centered radially by the roller bearing (11). Due to the kinematic system (3), the blades (2) perform an oscillating.

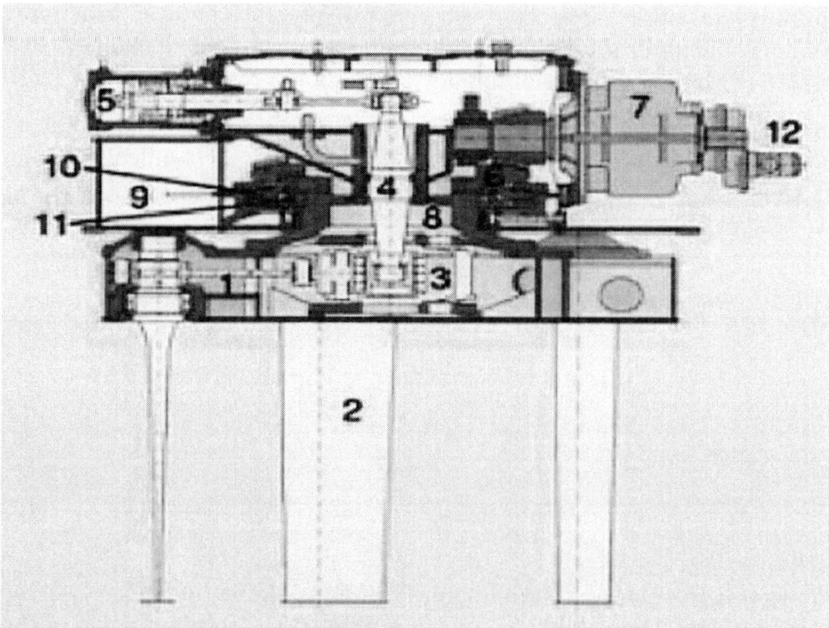

Figure 1. Components of VSP [5]

The amplitude and phase of blade motion is determined by the position of the steering center and hence the magnitude and direction of thrust are varied by means of the control rod [13, 14, 15].

The control rod is actuated by two orthogonally arranged servo motors (5). The propulsion servo motor (aservo motor is a rotary actuator that allows for precise control of angular position, velocity and acceleration) isused to adjust the pitch for longitudinal thrust (forward and reverse motion of the ship). The rudder servo motor is usedfor transverse thrust(motion to port and starboard). The two servo motorspermit steering according to Cartesian X/Y coordinates (identical with the principal axes of the ship). Controlled changes in thrust are possible via the thrust-free condition. For example, its direction can be changed from full ahead to full astern at a constant speed of rotation without creating disturbing transverse forces [16, 17].

Fitting the VSP with mechanical controls for the servo motors and an oil pump (12) flanged to the input gear results in a self-contained propulsion and maneuvering system which, apart from the torque input, requires no other source of power. Various mechanical kinematic systems and hydraulic blade actuating systems were developed and used in the course of development [18].

The blades of the VSP move along a circular path while simultaneously performing a superimposed pivoting motion. The perpendicular of the chord of the profiles intersects at single point, the steering center N.

3. GEOMETRICAL EQUATIONS IN VSP

The eccentricity (e) refers to as pitch of the VSP and is defined as:

$$e = \frac{ON}{D/2}, \quad (1)$$

where D is the blade orbit diameter.PointN, as shown in Figure 2, is the control point of the all blades which is related to the pitch angle of the VSP.

The advance coefficient (λ) of the VSP is the ratio of inflow velocity of the propeller (V_A) to the circumferential velocity u of the blades (u):

$$\lambda = \frac{V_A}{u} \tag{2}$$

The circumferential velocity u at the blade circle, with rotor speed (n) and blade orbit diameter (D), is given by

$$u = \pi n D \tag{3}$$

The motion of the blade relative to a stationary observer arises from the superimposition of the rotary movement and a straight line representing the forward motion of the vessel. The blade follows a curve of a cycloid. The rolling radius of the cycloid is $\lambda \cdot D/2$. In one revolution, the propeller travels a distance $\lambda \cdot D/2 \cdot \pi$ in the direction of motion of the ship. Because the blades travel along a cycloidal path, the VSP is also referred to as cycloidal propeller [19].

There is continuous variation in the lift during each revolution owing to the non-stationary inflow to the propeller blades. The force-components acting transversely to the desired direction of thrust cancel each other out, while the force-components in the direction of thrust add up over the circumference of propeller [7].

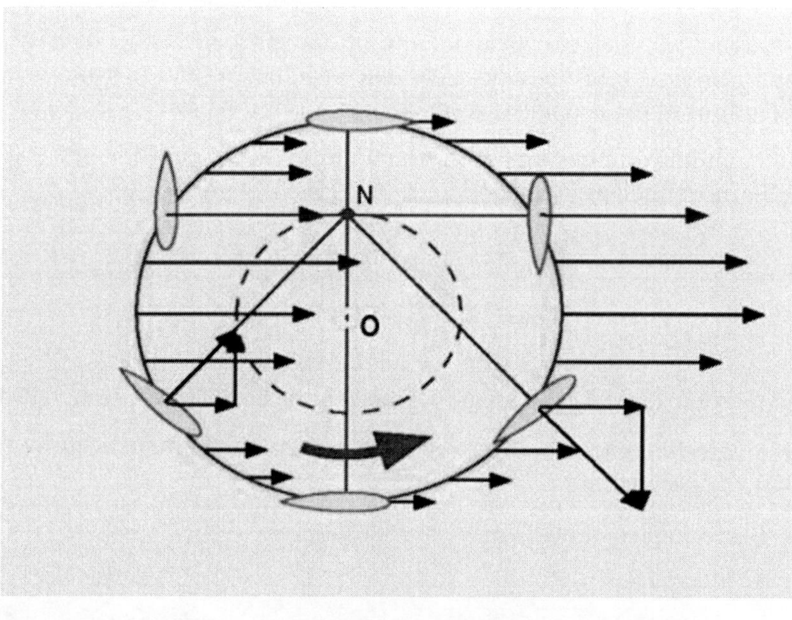

Figure 2. Forces on the blade in various angular positions [7]

4. HYDRODYNAMIC CHARACTERISTIC OF VSP

The hydrodynamic characteristics of VSP are represented by dimensionless coefficients which, as a result of the historical development of the system, differ by constant factors from those of screw propeller [20]. Hydrodynamic equations of the VSP are defined as follows:

$$\text{Thrust} \qquad T = \sum L_i \tag{4}$$

$$\text{Torque} \qquad M = \sum D_i \tag{5}$$

$$\text{Thrust coefficient} \quad K_S = \frac{T}{0.5\rho DLV^2} \tag{6}$$

$$\text{Torque coefficient} \quad K_D = \frac{4M}{\rho D^2 LV} \tag{7}$$

$$\text{Open - water efficiency}\ \eta_0 = \lambda \frac{K_S}{K_D} \tag{8}$$

where D, ρ, L, λ, n are the diameter of the propeller, the density of water, the length of the blades, advance coefficient and the number of revolutions of the propeller, respectively.

The Reynolds number for VSPs, based on the mean chord length c of a profile, is defined as follows [12]:

$$\mathrm{Re} = \frac{c}{\nu}\sqrt{V_A^2 + u^2} \tag{9}$$

where ν and c are the kinematic viscosity and chord length, respectively.

The hydrodynamic properties of VSPs are primarily influenced by the following parameters:

- Blade angle of attack during revolution.
- Profile geometry.
- Ratio of chord length c to blade circle diameter D, (c/D).
- Relative thickness of the profile.

- Blade shaft position.
- Ratio of blade length L to blade orbit diameter D (L/D).
- Shape of blade outline.
- Design of the blade ends [14-18][14].

The thrust coefficient and efficiency are a function of the advance coefficient λ and pitch e. there is a qualitative shift in the coefficients as a function of λ and e, as is familiar from screw propellers with variable pitch. The efficiency increases with pitch. In addition, the fewer the number of blades has higher efficiency. On today's VSPs, the number of blades is much less significant since the hydrodynamic interactions between the blades on the front and rear halves of the rotor are reduced by modification of the blade angle curve.

The side thrust of VSP increases with the advance coefficient. As the advance ratio increases and the propeller speed remains the same, the steering forces also increase, as do the torques [7].

5. PRINCIPAL TUG PARTICULARS

The main particulars of a tug are shown in Table 1. This will be usedfor estimateof the lift, drag, pressure and velocity distributions in VSPs.

Table 1. Principal tug particulars

	Parameter	Value [unit]
Length of waterline	LWL	14 [m]
Beam	B	3.3 [m]
Draft	T_l	0.9 [m]
Block coefficient	C_B	0.41 [-]
Speed	V	12 [Knot]
Total resistance	R_T	13.29 [kN]

The present VSP has 4 blades and the blade orbit diameter is 1.5 m. The rotor speed is 150rpm. NACA 0012 blade section is selected that has a symmetric section.

The blade dimensions are defined:

Chord length= 1 m, Span length= 1.2 m, Relative thickness = 0.02 m.

6. NUMERICAL RESULTS

This study used the finite volume code FLUENT v14 to solve the incompressible continuity and momentum equation in two dimensions. For the turbulent flow cases, the equations were Reynolds-averaged and the $k-\varepsilon$ model was used for turbulence closure. This study adopted the re-normalization group (RNG) extension of the $k-\varepsilon$ model.

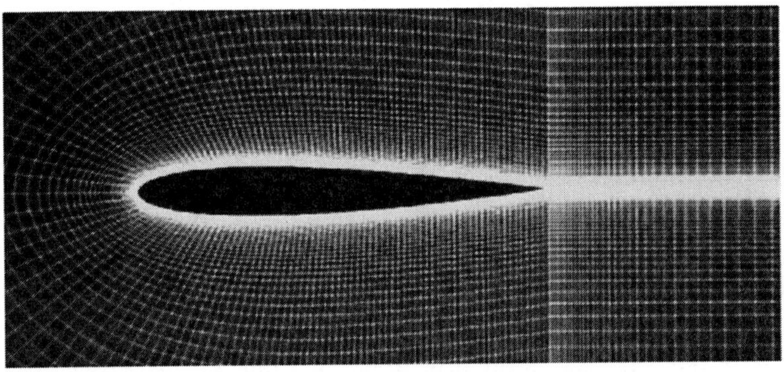

Figure 3. Mesh and details of the profile mesh

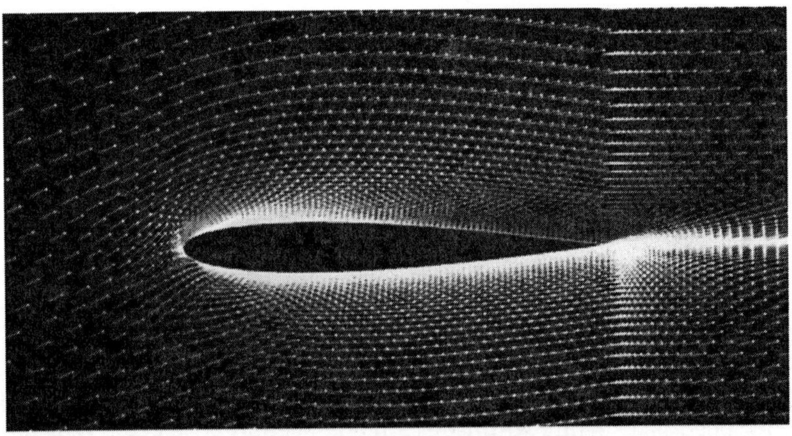

Figure 4. Velocity vector around the blade of the VSP at α=15°

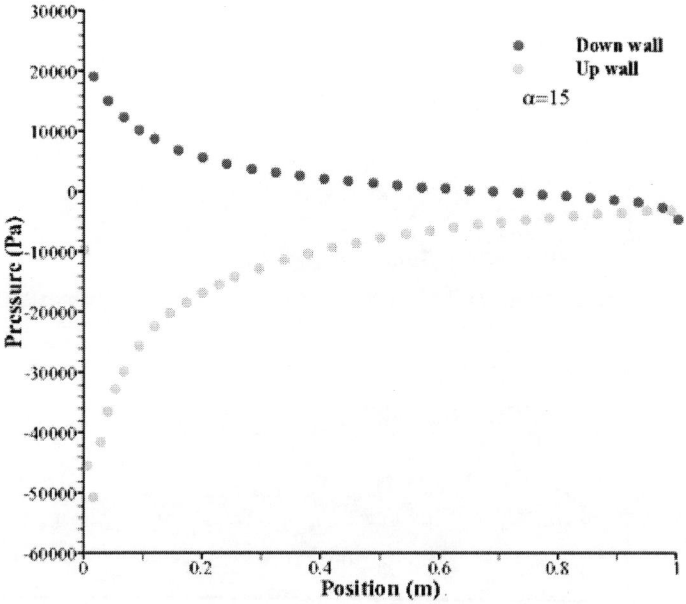

Figure 5. Pressure distribution on blade at α=15°

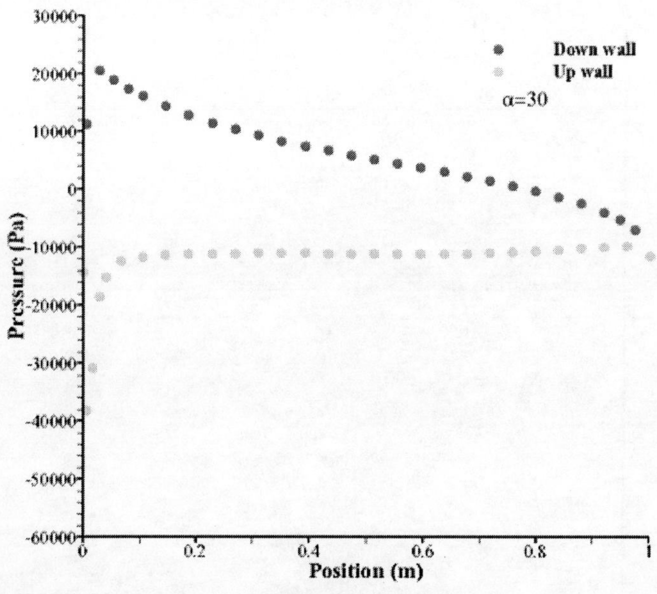

Figure 6. Pressure distribution on blade at α=30°

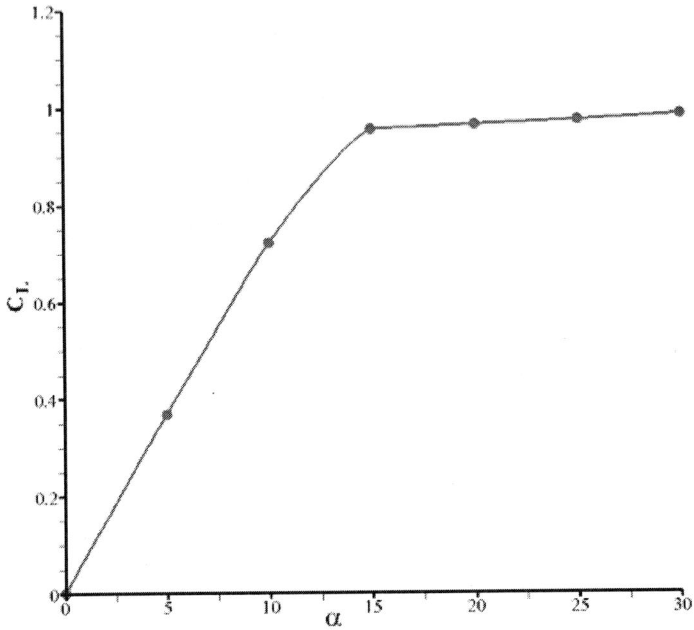

Figure 7. Lift coefficients as a function of α

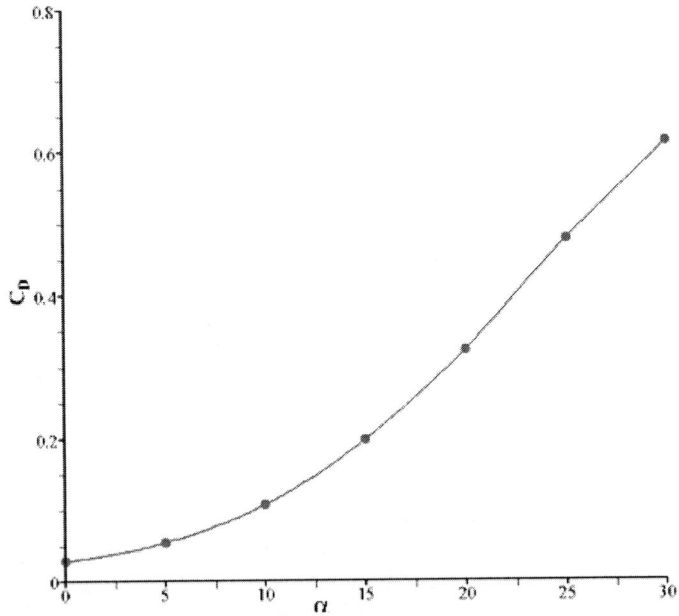

Figure 8. Drag coefficients as a function of α

Here, the numerical results of the blade of VSP are presented. The results are velocity vector, pressure distributions on blade of VSP, lift and drag at various angles of attack (α). Figure 3 shows the mesh and its domain. Numerical results for pressure, velocity, lift and drag are shown in Figure 4-Figure 8.

7. HYDRODYNAMIC CHARACTERISTIC RESULTS

The most important of VSP characteristics of the VSP are hydrodynamic coefficients. These coefficients are thrust, moment and efficiency. The VSP is a variable pitch propeller. These calculations are carried out for various pitches 0.3, 0.4 and 0.5. Hydrodynamic characteristics (thrust, moment and efficiency) are shown in Figure 9-Figure 11. The efficiency increases with thepitch. Advance velocity coefficient (λ) can play an important role in the hydrodynamic performance. Maximum efficiency is around 70% for pitch ratio of 0.5 at $\lambda = 0.7$ and hasreasonable value based on the experimental data as shown in reference [5].

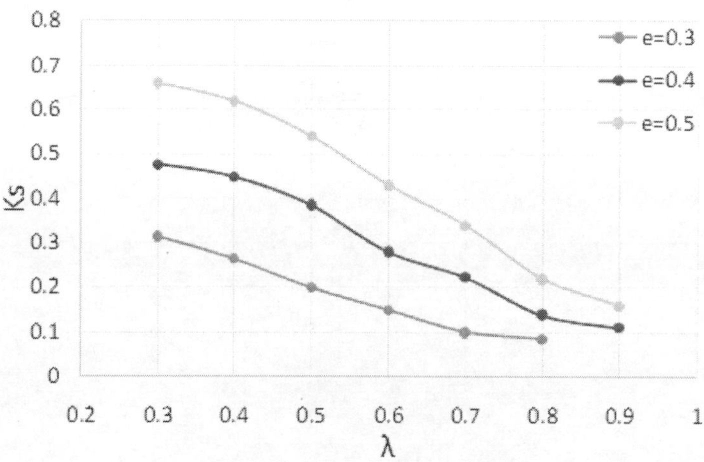

Figure 9. Thrust coefficient at different pitch ratios

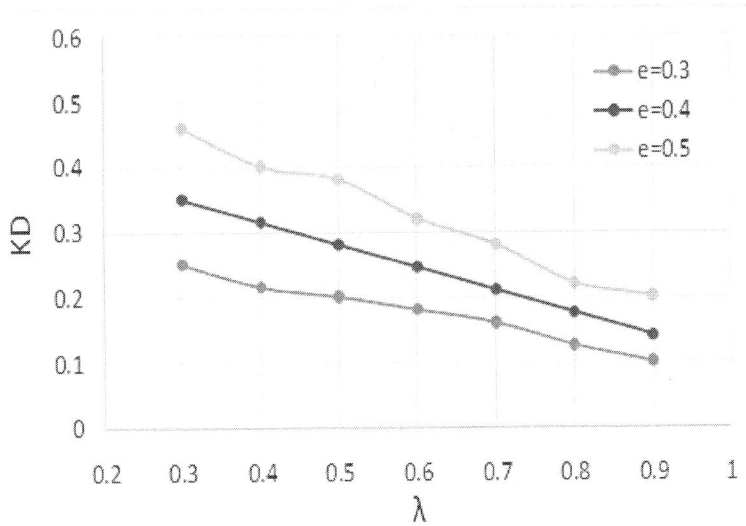

Figure 10. Torque coefficient at different pitch ratios

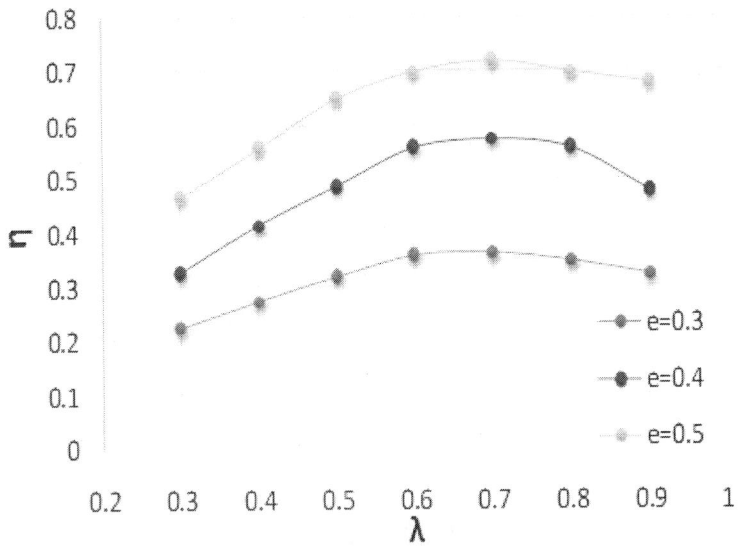

Figure 11. Efficiency at different pitch ratios

8. CONCLUSIONS

The hydrodynamic analysis of a VSP at various operating conditions was investigated. A finite volume–based RANS solver (Fluent) has been used to evaluate the performance of these systems. Based on the numerical results, the following conclusions can be drawn:

1. When the point N (means control point) is changed the resultant velocity entering to the blade and its angle are determined. Therefore, especial emphasis should be applied to determine the physical and geometrical blades operating conditions.
2. Physical characteristics of one blade at various angles of attack is investigated in order to get the lift and drag of the blade.
3. Hydrodynamic characteristics of the VSP(K_S, K_D, η) are calculated at three different pitch rations. It is concluded that with increasing the pitch of the blade, efficiency is increased.

ACKNOWLEDGEMENT

This research was supported by the Marine Research Center of Amirkabir University of Technology whose works are greatly acknowledged.

NOMENCLATURE

B	Beam, m
c	Chord length, m
C_B	Block coefficient
C_L	Lift coefficient
C_D	Drag coefficient
D	Blade orbit diameter, m
e	Pitch
K_s	Thrust coefficient
K_D	Torque coefficient
L	length of the blades, m
LWL	Length of waterline, m
M	Torque, kN
n	Propeller rotational speed, rps
N	Steering center
O	Center of VSPs
Re	Reynolds number
R_T	Total resistance, kN
T	Thrust, kN
T_1	Draft, m
u	Circumferential velocity, ms^{-1}
V	Speed of ship, Knot
V_A	Inflow velocity of propeller, ms^{-1}
λ	Advance coefficient
ν	Kinematic viscosity, $m^2 s^{-1}$
α	angles of attack, degree
ρ	Density of water, $kg\ m^{-3}$

REFERENCES

1. www.en.wikipedia.org/wiki/Voith_Schneider_Propeller.
2. www.voithturbo.com/marine.
3. Dirk Jürgens, Michael Palm, "VSP - An Efficient Propulsion System for DP Controlled Vessels", Voith Turbo Schneider Propulsions, October 2009.
4. Dirk Jurgens, Michael Palm, Sebastian Singer, Karsten Urban, "Numerical optimization of the Voith-Schneider Propeller", Journal of Applied Mathematics and Mechanics, October 2007.
5. Jens-Erk Bartels, Dirk Jürgens, The Voith Schneider Propeller: Current Applications and New Developments,Voith Turbo Marine GmbH & Company KG, 2006

6. www.voith-marinetechnology.com.

7. Dirk Jürgens, Torsten Moltrecht, Marcel Flipse, "Cycloidal propulsion the Quiet Maneuvering propulsion for Large Motor Yachts", Paper to 18th International HISWA, Symposium on Yacht Design and yacht Construction,November 15- 16, 2004

8. www.airfoiltools.com/airfoil/details?airfoil=n0012-il.

9. Jürgens, Birgit; Fork, Faszination, "Voith Schneider propeller", Geschichte und technik, Humburg 2002.

10. Kirsten, F.K., The Cycloidal propeller in marine Application. University of Washington , 1950

11. Coton, F.N, Galbraith, A.McD, "An Experiment study of dynamic stall on a finite wing". Aeronautical Journal, p.229-236, may, 1999

12. Ficken, N.L, Dickerson, M.C., "Experimental performance and steering Characteristic of cycloidal propeller",NSRDC report 2983, Washington, 1969

13. Zhu, D.M., "A Computational Method for Cycloidal propeller". I.S.P., vol.28, No.321, 1981.

14. Isay, W.H., "Moderene probleme der Propeller theorie". Berlin, Heidelberg, Springer-Verlag London Ltd, New York 1970.

15. Streckwall, H.:An unsteady "Vortex-lattice method for ships' propeller", shipbuilding symposium in rostock, 1991.

16. Ferziger, J, H. Peric, M., Computational methods for fluid dynamics, Springer, Third edition , Berlin, Heidelberg, 2013.

17. Bartels, J.-E. "Voith – Schneider propulsion systems for double ended ferries", Log's ferries 93 conference, Florida, .march 1993

18. Jürgens, D, "Theoretical and experimental studies of unsteady flows around blades and development of calculation method for vertical –axis rotors", Dissertation, Rostock University 1994

19. Sturmhoefel, U; Bartels, J.-E.: Voith, J. M, "Basic requirements for safe escort vessels-theoretical consideration and model measurements", RINA, Escort Tugs design, construction and handling-the way ahead".London, 1993.

20. Mott, H.J, Kirstensen, D.H. laible, D.H., Thompson, T.,Stone ,M.; "Acquisition of tanker escort and assist tugs for Newfoundland Transshipment", SNAME, 1998.

21. Ralf Leidenberger, Karsten Urban, "Automatic differentiation for the optimization of a ship propulsion and steering system: a proof of concept", Springer Science+Business Media, LLC, 2010

22. Piergiorgio Pasetto, Leslie Barnabà, Matteo Miletto Bracco, Igor Zotti, "A NEW PROPOSAL OF A VERTICAL AXIS PROPELLER", International Conference IDS2013 – Amazonia, July 2013.

11
CHAPTER

3D Numerical Simulation of Propeller and its Aerodynamic Interference Effects on Tail of a Flying Boat

Hossein Afshar and Mohammad Ali Keshvari

Department of Mechanical Engineering, East Tehran Branch, Islamic Azad University, Tehran, Iran

ABSTRACT

In this study, considering the complexity of flow and its three dimensional nature, aerodynamic coefficients of a propeller is obtained and the effects of propeller flow on the tail were analyzed by solving Navier-Stokes equations numerically. The results indicated that the position of the propeller toward the tail has to neutralize the torque produced by the propeller as much as possible. At the same time, the lift and drag of the tail have to be in the proper range. The best angle and direction for propeller installation is proposed. Computational fluid dynamics were applied to find the flow about the propeller and the tail and to determine their interaction effects.

INTRODUCTION

Determination of the interference effects of a flying object component is crucial in designing and optimizing its complex shape. An appropriate design will require an arrangement of components that result in positive interference effects. Such effects can only be obtained by evaluating each component once alone and once in combination with other components through simple, cost-effective methods of computational fluid dynamics.

A propeller uses the engine power and creates thrust whose value is not constant for a particular propeller, i.e., it depends on the propeller's diameter, linear velocity and rotational speed. Although decreased inlet flow rate is generally associated with increased thrust, considering the definition of relative velocity, the geometry of a propeller and thus flow separation (stall) in some parts of the propeller blades, especially in those with great airfoil angles at the root, reduces the static thrust when reaching inertia. Therefore, static thrust is not an appropriate criterion for propeller assessment and the propeller is designed for maximum productivity at a certain speed.

Meanwhile, propellers with similar shape and equal form factors will display similar behaviors. Performance of propellers is assessed based on a number of dimensionless coefficients including advance ratio, power factor and thrust coefficient. Moreover, in order to evaluate the effects of the propeller on the tail, it is first necessary to somehow determine the propeller flow field.

There are many studies related to WIG's operation. Some of them are Joh and Kim (2004) numerically studied the pressure changes, drag and static height stability a round WIG craft airfoil two and three dimensional configuration. Kim *et al.* (2009) performed the optimization of the shape and the configuration of the WIG craft including fuselage, main wing, tail and side wing, The SQP (Sequential Quadratic Programming) for the optimization and the VLM (Vortex Lattice Method) for the evaluation of the flow around a WIG craft were employed. Kornev and Matveev (2003) analyzed the static height stability using the vortex lattices method. Rozhdestvensky (2006) attempts to improve the stability and controllability of a WIG effect vehicle have included employing a large horizontal tail, a tandem configuration. Afshar and Alishahi (2007) the flow field about a complete flying-boat in ground effect is resolved. The influences of using winglet in ground and out of ground effect are compared and it is shown that ground effects this influence to a large extent. Park and Lee (2008) carried out a numerical investigation on the effect of an endplate at various angels of attack and ground clearances. They found that the end plate prevented the high pressure air form escaping out of the lower wing surface and so reduced the influence of the wing-tip vortex and further argument the lift and drag ratio. Kikuchi *et al.* (1997) measured the interaction forces on the air foil and compare their result with those of the numerical results. Yakhot *et al.* (1992) described that it's known that the RNG k-ε model with an additional term in its ε-equation can significantly improve the accuracy of turbulent flow. Chun and Chang (2002) performed the experimental study for the static and dynamic stability based on the wind tunnel test. The model vehicle was the 20-passenger WIG craft. They clearly showed the implication of the static height and dynamic stability in their study. They concluded that one of the ways in increasing the static stability and the damping ratio for the short period pitching oscillation was to increase the tail wing size. However, they pointed out that the large tail increased the structural weight and the drag. Kim and Chun (1998) performed computational optimization for an airfoil shape. They chose the pressure distributions (inverse design) and lift coefficient as the objective functions and obtained their optimal solutions using a Sequential Quadratic Programming (SQP) method which, is a gradient-based local optimization technique. Lee and Lee (2011) performed a numerical optimization on a three-dimensional wing in ground effect. They classified four different groups of pure to individuals according to the wing configuration: The drag and static height stability group, the lift group and the intermediate group. Hirata and Hino (1997) conducted a numerical investigation on Power Augmented Ram (PAR) wing in ground effect, considering two boundary conditions: (1) A moving belt ground plate condition and (2) A fixed ground plate condition corresponding to a wind tunnel test.Murao *et al.* (2005) stated that two ducted propellers were installed before the wing. The lift force is increased by 50% as compared to that without PAR. Wind tunnel performance of a WIG craft with Propeller-Deflected Slipstream (PDS). In calculating aerodynamic forces on lifting surfaces of WIG craft in cruising flight regimes, the water surface deformations are usually minimal and neglected (Barber *et al.*, 1998; Masuda and Suzuki, 1991). In calculating aerodynamic forces on lifting surfaces of WIG craft in cruising flight regimes, the water surface deformations are usually minimal and neglected. Later Molina and Zhang (2011) described when the airfoil is in close proximity to the ground, the flows are dominated by the viscous effects at the low reduced frequencies, whereas the inviscid flow behaviors are acceptable at the high reduced frequencies. Tuck (1984) the water surface deformations can become significant and change the effective dimensions of the channel between the wing and water, thus affecting the aerodynamic forces. For steady analysis of this flow with water surface deformations, a hydrostatic assumption the water surface depression (Ellington *et al.*, 1996) via a classical linear theory about a vertically oscillating wing in the inviscid flow.

METHODOLOGY

Geometry of the propeller: An airplane propeller with 10 ft diameter and a fixed pitch of 7 ft was selected to assess the effects of propeller on tail. The physical and aerodynamic twist of the propeller and it's taper from root to tip is shown in Fig. 1. The propeller is divided into subsections for structured grid generation as O type for each blade and H type for the space between two blades.

The airfoils are also presented at different radius in Fig. 2. The dimensions of all figures are expressed in mm.

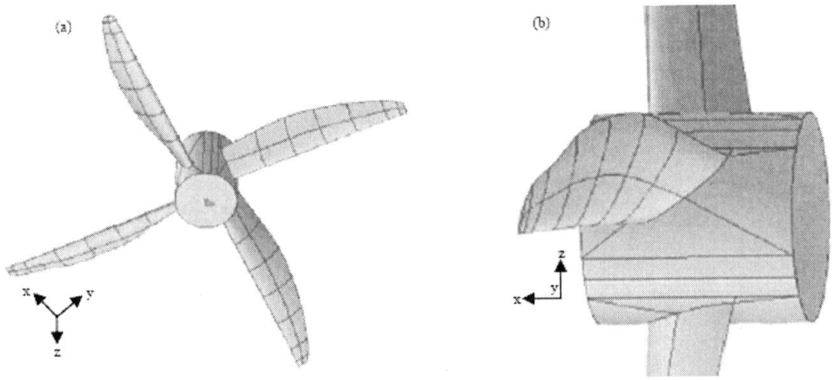

Figure 1(a-b): Physical and aerodynamic twist of the propellers

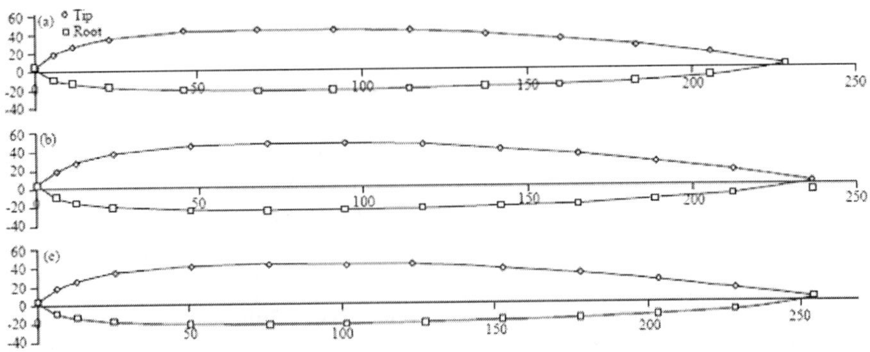

Figure 2(a-c). Dimensions of airfoils, (a) r/R = 0.15, (b) r/R = 0.3 and (c) r/R = 0.45

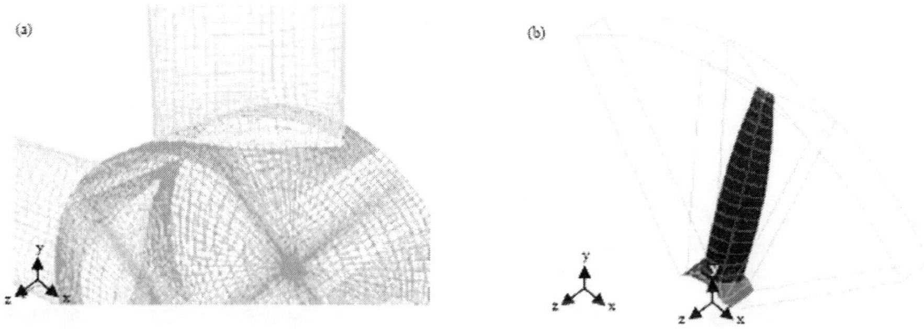

Fig. 3(a-b): Grid generation around the propeller

The airfoil installation angle is 56 and 12.5° at root and tip, respectively. Besides, its chord is 228.5 and 67.5 mm at root and tip, respectively.

Grid generation around the propeller: Physical and aerodynamic twist of the blades and varying chord and thickness of the airfoil from the root to the tip complicate the process of grid generating around the propeller. Hence, after testing all possible grids, a structured grid comprising O and H grids (with some unstructured parts) was created. As shown in Fig. 3, the O and H grids were used around each blade and between two blades, respectively. Grids on the front and rear plates of the hub and on the tips were unstructured. Grids on the front and rear plates (a little before and after the propeller) were in the form of O type.

Governing equations: Since the Mach number of boat speed was less than 0.3, the flow was considered to be incompressible. The Reynolds number was computed at about 10 and the flow was thus turbulent. The terms resulting from turbulence in the transport equation were modeled with the k-ε turbulence model. Non-temporal governing equations were as follows.

The continuity equation:

$$\frac{\partial u}{\partial x}+\frac{\partial \upsilon}{\partial y}+\frac{\partial \omega}{\partial z}=0$$

Navier-Stokes equation in the inertial frame:

$$\rho\nabla.(\bar{\upsilon}.\bar{\upsilon})=-\nabla p+\nabla.(\bar{\bar{\tau}})$$

Navier-Stokes equation in the rotating frame with absolute velocity:

$$\nabla.(\rho\bar{\upsilon}_r\,\bar{\upsilon})+\bar{\Omega}\times\bar{\upsilon}=-\nabla_p+\nabla.\bar{\bar{\tau}}$$

where, u, υ and ω are the components of velocity along the x, y and z axes, respectively.

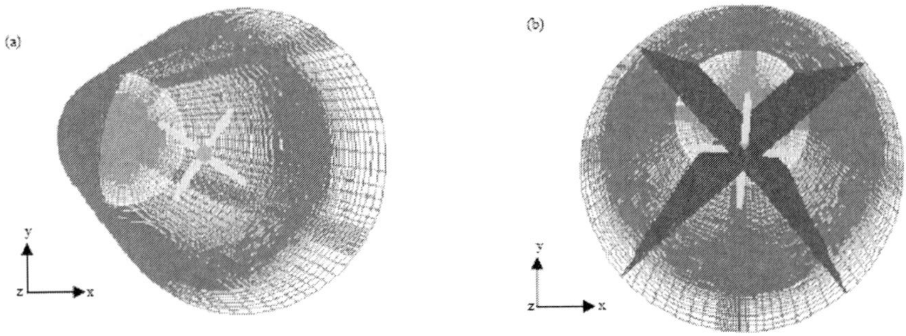

Figure 4(a-b): Propellers in rotating coordinate system

Moreover, τ_{ij} is the components of stress tensor and is defined as:

$$\bar{\bar{\tau}}=(\mu+\mu T)\left[(\nabla\bar{\upsilon}+\nabla\bar{\upsilon}^{T})-\frac{2}{3}\nabla.\bar{\upsilon}I\right]$$

where, $\bar{\upsilon}$ is the velocity vector and I is the unit tensor.

The k-ε equations:

$$\rho\frac{\partial}{\partial_{xi}}(k\upsilon_i)=\frac{\partial}{\partial_{xj}}\left[\left(\mu+\frac{\mu_t}{\sigma_k}\right)\frac{\partial_k}{\partial_{xj}}\right]+G-\rho\varepsilon$$

$$\rho\frac{\partial}{\partial x_i}(\varepsilon u_i)=\frac{\partial}{\partial x_j}\left[\left(\mu+\frac{\mu_t}{\sigma_\varepsilon}\right)\frac{\partial_\varepsilon}{\partial x_j}\right]+C_1\frac{\varepsilon}{k}G-C_2\rho\frac{\varepsilon^2}{k}$$

$$\mu_t=\rho C_\mu\frac{k^2}{\varepsilon}$$

where, C_1 and C_2 are empirical coefficients and σ_ε and σ_k are the turbulent Schmidt and Prandtl numbers:

$$C_1 = 1.44, C_2 = 1.92, C_\mu = 0.09, \sigma_k = 1, \sigma_\varepsilon = 1.3$$

Boundary conditions: The boat's speed and the propeller's angular velocity are considered as 30 m sec^{-1} and 237 rad sec^{-1}, respectively. The rotational flow produced by the propeller-a temporal flow that depends on the number of blades (n = 4 in this study)-destroys the symmetry.

Considering the propeller inflow and tip Mach numbers (<0.3 and about 0.62, respectively), the flow is compressible at the tips. Nevertheless as the region of compressible flow is too small compared to the evaluated domain, the flow was assumed to be incompressible and the equations were thus elliptic. Supposing the air to move toward the object, the velocity was set at the front, side, bottom and top plates as boundary condition. At the output plate, however, the boundary condition was determined as the pressure. A no slip condition was also adopted on the blades and at the horizontal and vertical tails. Knowing that only the propeller rotates and other parts are fixed, the air around the propeller is separated from the rest of the domain and the flow equations in this region are solved in a rotating coordinate system.

Solution: The computational domain for obtaining aerodynamic characteristics of the propeller is shown in Fig. 4. Finite-volume method was used to solve the governing equations. The SIMPLE algorithm with first-order accuracy was applied to couple velocity and pressure. Momentum and flow-related equations were solved using a first-order upwind scheme. Turbulence was modeled with the standard k-ε turbulence model.

The volumes around the objects were assessed in grids containing 6×10^5, 1.55×10^6, 2.1×10^6 and 3.0×10^6 cells. While the results from the first two grids had substantial differences, the two latter grids were slightly different. In fact, using the k-ε turbulence model has moved the nodes in the three-million cell grid toward the object's borders and placed y+between 30 and 60. Consequently, this grid was used in calculating the aerodynamic coefficients.

RESULTS

Pressure distribution along the propeller and the hub: Figure 5 and 6 demonstrate the pressure counters on the propeller blades and the hub. Maximum pressure was toward the leading edge and the rear part but dramatically reduced thereafter.

The pressure was maximum at the tip but decreased and so did the thrust, along the blade to the root. Changes in pressure on the hub were smaller than those on the tip. Hence, the hub has to be evaluated alone if pressure distribution is to be shown. Pressure distribution on the hub is shown in Fig. 7.

As the Fig. 7 depicts the cross-section of the airfoils on the hub, pressure was high in the back of the propeller and low on the airfoil, i.e., a driving force in the positive direction of the x-axis was expected.

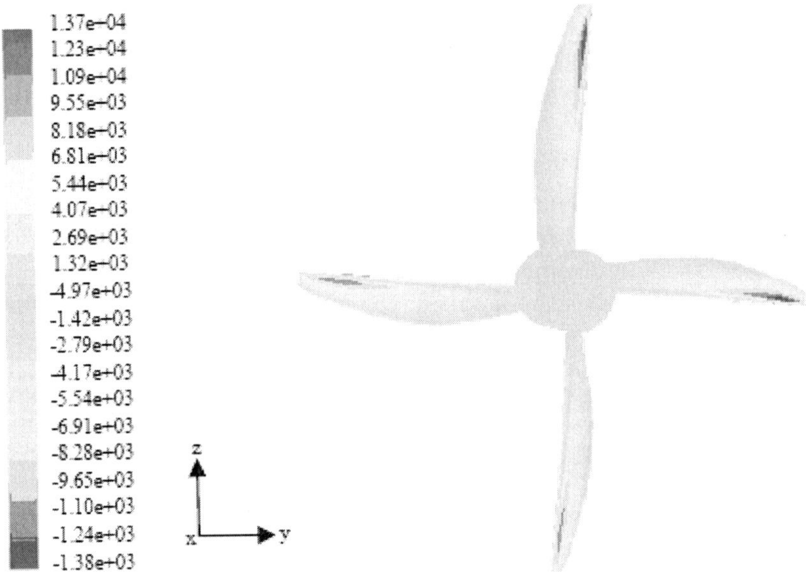

Figure 5: Pressure distribution on the front face of the propeller

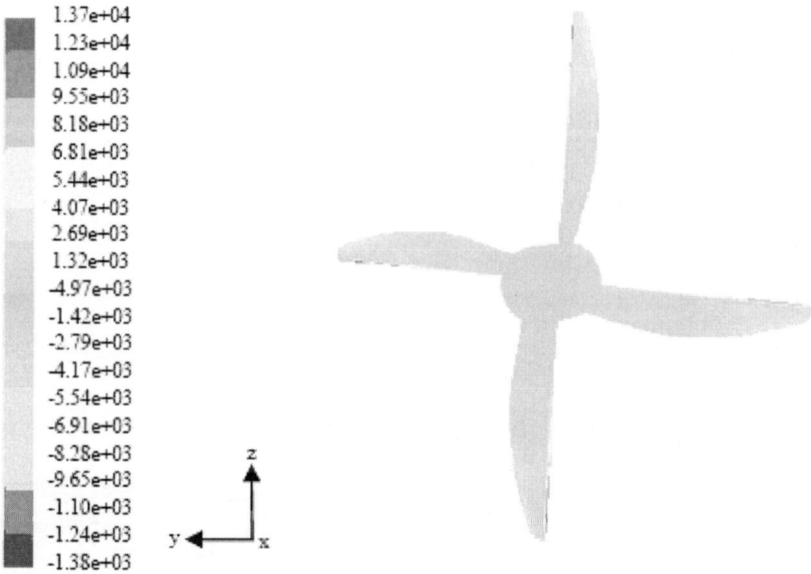

Figure 6: Pressure distribution on the rear face of the propeller

Pressure distribution on the tail: Pressure distribution on the tail was assessed in three states:

- There was no angle between the propeller and the junction of the two tails
- The propeller was 7° below and 2.5° to the left of the junction of the two tails and the boat moved in the direction of the propeller
- The propeller was 7° below and 2.5° to the left of the junction of the two tails and the boat moved in the direction of the tail

In the first state, none of the propeller and the tail was in the angle of attack. In the second state, the propeller was not at the angle of attack but the tail was. In the third state, however, the propeller was at the angle of attack but the tail was not.

The x-axis was along the longitudinal direction of the flying boat from the end to the nose. The z-axis was upward and the y-axis pointed to the left when standing in the direction of the x-axis. The propeller rotated in the negative direction of the x-axis. Hereafter, left and right are defined based on the positive direction of the x-axis. Since the propeller turned counterclockwise (a left handed propeller), the vortex caused by the propeller hit the top and left surfaces of the horizontal tail, the left surface of the vertical tail and the bottom and right surfaces of the horizontal tail and thus increased the pressure at these areas. Figure 8 shows the pressure distribution counter when the propeller had no angle with the junction of the two tails.

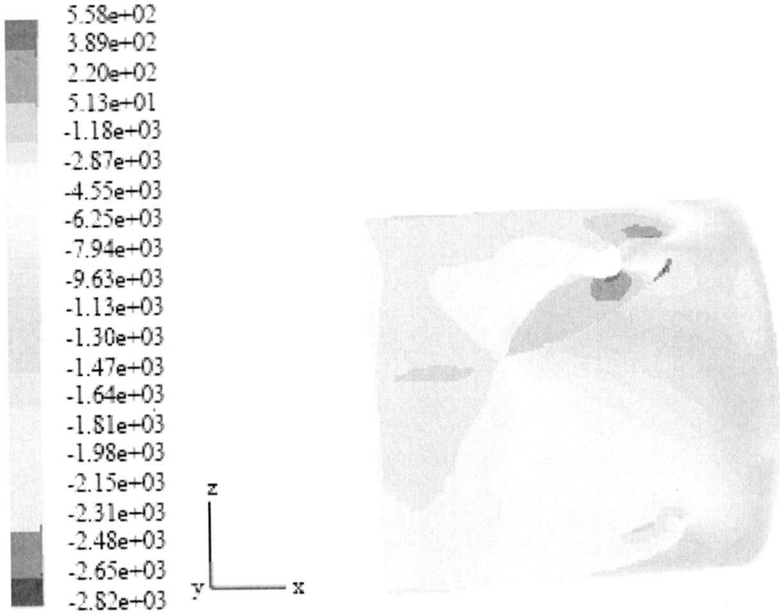

Figure 7: Pressure distribution on the hub

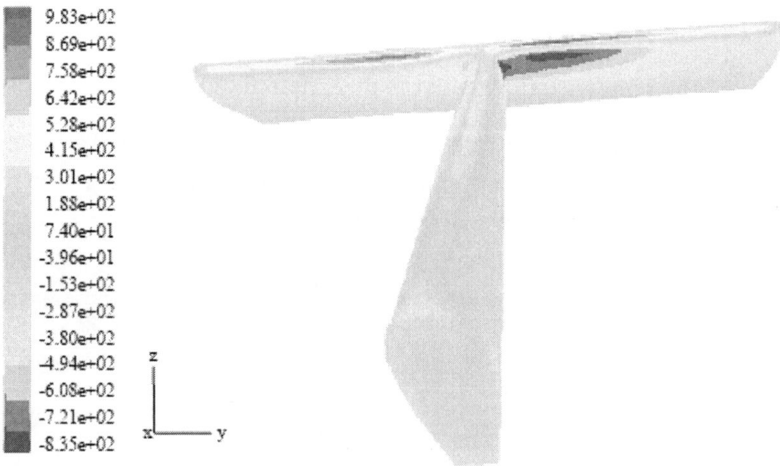

Figure 8: Pressure distribution counter when the propeller had no angle with the junction of the two tails

This pressure distribution would produce a torsional moment in the negative direction of the x-axis and also in the positive direction of the z-axis in the operating conditions of the propeller. Such a situation would create the yaw moment which can be neutralized by turning the propeller in a way that its axial flow hits the right side of the vertical tail. This justifies the 2.5° diversion of the propeller.

Figure 9 shows pressure distribution on top and bottom surfaces of the horizontal tail.

Evidently, maximum pressure was imposed on the right leading edge of the horizontal tail and the pressure distribution on the leading edge was completely non-uniform.

Figure 10 shows pressure distribution on the tail when the propeller made an angle with it and the boat moved in the direction of the tail.

As it is seen, the propeller shaft has diverted downward. It can thus be concluded that the vortex caused by the propeller has had the least contact with the horizontal tail. Therefore, the minimum pressure underneath the horizontal tail has increased compared to the conditions when propeller shaft was not diverted.

Velocity distribution: Velocity distribution at different parts of the solution domain was studied when the propeller and the tail made an angle and the boat was moving in the direction of the tail.

Figure 11 demonstrates velocity vectors along the propeller blades.

Considering the small turning radius near the root, the tangential velocity of the flow was also small at this area.

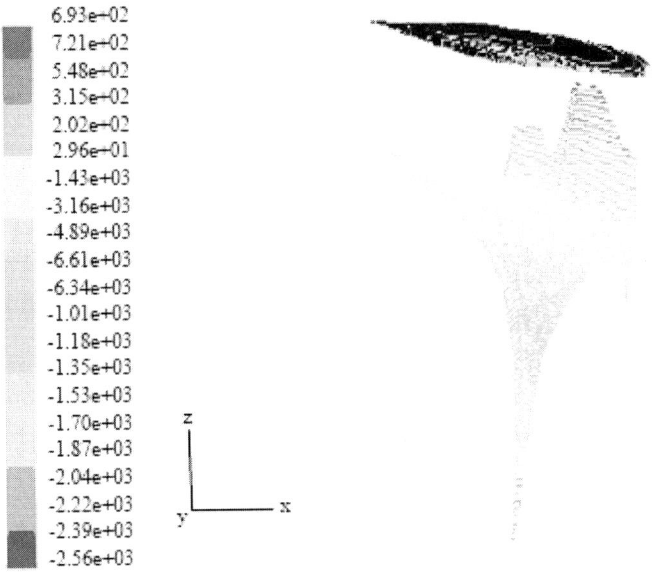

Figure 9: Pressure distribution on top and bottom surfaces of the horizontal tail

As the angle between the chord line and the airfoil at the root was about 56°, the angle of attack caused by the relative wind was a high, negative value at the root. In other words as the Fig. 10 displays, the flow separated near the root and an adverse pressure gradient occurred. Due to the vortexes produced at the tip, maximum velocity was detected at a radius equal to 0.75 of the propeller radius. The obtained maximum velocity (122 m sec^{-1}) and Mach number (0.37) under such conditions prove the assumption of incompressible flow to be true.

Figure 12 shows the axial velocity distribution on the propeller plate.

Apparently, the axial velocity distribution on the propeller was absolutely uniform. On the blades, however, it was substantially great compared to the distance between two blades. In addition, axial velocity on each blade decreased from the tip to the root

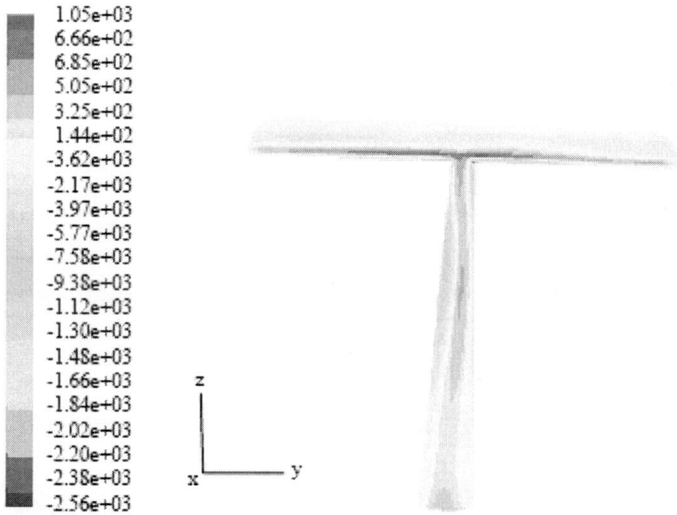

Fig. 10: Pressure distribution on the tail when the propeller made an angle with tail

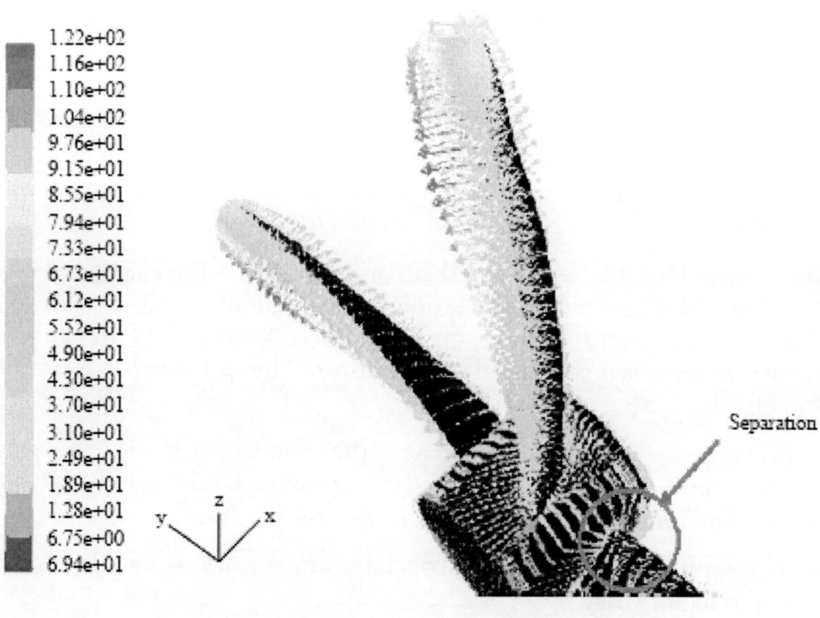

Fig. 11: Velocity vectors along the propeller blades

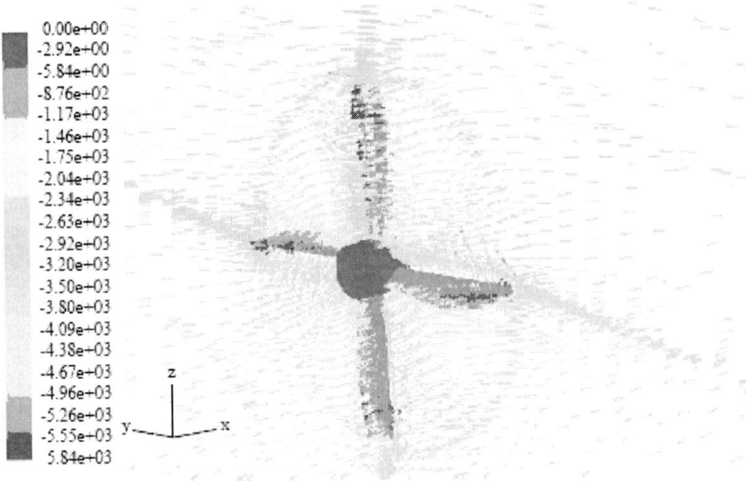

Fig. 12: Axial velocity distribution on the propeller plate

DISCUSSION

The thrust of the propeller under static operating conditions is measured and it is compared with numerical results. The comparison for different operating conditions is shown in Table 1.

In all conditions, the measured thrust is higher than the calculated one. As a fixed-pitch propeller was studied, the angle of airfoils at different cross-sections could be measured which is summarized in Table 2. These values were compared with those obtained from the AutoCAD model.

Apparently, the error was higher near the root. Under static conditions and low rotation, the thrust is predominantly generated by the airfoils close to the root.

Forces applied to the tail and propellers under various conditions are summarized in Table 3.

Table 1: Measured thrust of the propeller, J = v/n.D

Propeller-fix pitch = 58							
n (rpm)	n (rad sec^{-1})	D (m)	V (m sec^{-1})	J	Thrust (kg)	Thrust (kg) (Experiment)	Error (%)
1200	51.71	1.62	0.1	0.0075	9.387755102	29	67.628
3000	129.28	1.62	0.1	0.003	59.02040816	70	15.685
4000	172.38	1.62	0.1	0.00225	105.1020408	110	4.4527
5000	215.47	1.62	0.1	0.0018	164.1959184	182	9.7825
5200	224.10	1.62	0.1	0.001730769	177.8448980	192	7.3724
5500	237.00	1.62	0.1	0.001636364	198.7387755	230	13.592

Table 2: Parameters of the propeller

r/R	Blade angle according to pitch = 58"	Measure blade angle in AutoCAD	Error (%)
0.192	55.69042639	37.40	32.843
0.308	42.37492325	34.20	19.292
0.423	33.59261509	30.00	10.695
0.538	27.57348956	26.10	5.3439
0.654	23.23998459	22.60	2.7538
0.75	20.53333816	20.20	1.6234
0.769	20.06287557	19.70	1.8087
0.884	17.62536715	16.02	8.0870

Table 3: Forces applied to the tail and propellers

Parameters	Thrust (N)	Mx (N.M)	Lift (N)	Drag (N) All drag numbers are positive	Perpendicular force to the horizontal tail for Y	M (V.H tail cross) (N.M)		
At FSV: 30 m sec^{-1}, PAV: 237 rad sec^{-1} and AAPT: 0°								
Propeller	606.15	285.16						
Propeller and tail with aerodynamic interferences effect	610.73	285.84	-68.03	-101.34	-125.28	-240.28	33.13	-162.06
At FSV: 30 m sec^{-1} and PAV: 0 rad sec^{-1}								
Propeller and tail with aerodynamic interferences effect	-243		-23.71	-87.88	30.14	48.49	36.98	36.24
At FSV: 30 m sec^{-1}, PAV: 237 rad sec^{-1} and Propeller position*								
Propeller and tail with aerodynamic interferences effect	626.77	287.36	-559.08	174.28	-131.94			
At FSV: 30 m sec^{-1}, PAV: 237 rad sec^{-1} and Propeller position*								
Propeller and tail with aerodynamic interferences effect	612.66	286.12	1037.05	-216.566	218.0075			

Drag of a typical flying boat which is investigated by Afshar and Alishahi (2007) is in the range of the present propeller thrust at 5500 rpm. So this propeller with the suggested installing condition can be used to power the flying boat.

CONCLUSION

Turning the propeller shaft 7° downward and 2.5° toward the negative direction of the z-axis when the boat is moving in the direction of the tail will place the propeller at the angle of attack. Since the angle of attack of the tail is zero in this situation, its drag will increase by 50%. In fact, even when the installation angle of the propeller is zero and the tail is at the angle of attack with the axial velocity component of the propeller flow, the angle of attack of the tail will not be zero due to the vortex produced by the propeller. Meanwhile, dramatic increase of the angle of attack in the second situation will definitely increase the torque about the z-axis.

When the boat moves in the direction of propeller shaft (which is generally the case since the thrust of the propeller as the only propulsion power of the flying boat, is in the direction of the propeller shaft), the applied negative lift force on the tail would be 14 times higher compared to the previous conditions. This state is appropriate for adjustments. Under such conditions, the torque would be negative and 70% less in value compared to the previous state. Therefore, turning the propeller by less than 2.5° will not only neutralize the torque but also reduce the boat's deviation from the x-axis.

The presented results show that the static balancing of a flying boat cannot guarantee its stability and there are restrict limitations in this regard. Considering the aerodynamic interaction of propeller with parts of a flying boat is mandatory in assembling the parts.

REFERTESNCES

1. Afshar, H. and M.M. Alishahi, 2009. A study of winglet and aerodynamics interference in 3-D viscous flow around a flying-boat in ground effect. J. Applied Sci., 9: 3752-3757.
2. Barber, T., E. Leonardi and D. Archer, 1998. Free surface deformation caused by a wing in ground effect over water.

Proceedings of the International Workshop on WISE Up to Ekranoplan GEMs, June 15-16, 1998, The University of New South Wales, Sydney, Australia -.

3. Chun, H.H. and C.H. Chang, 2002. Longitudinal stability and dynamic motions of a small passenger WIG craft. Ocean Eng., 29: 1145-1162.

4. Ellington, C.P., C. van den Berg, A.P. Willmott and A.L.R. Thomas, 1996. Leading-edge vortices in insect flight. Nature, 384: 626-630.

5. Hirata, N. and T. Hino, 1997. Investigation of a three-dimensional power-augmented ram wing in ground effect. Proceedings of 35th Aerospace Sciences Meeting and Exhibition, January 6-9, 1997, Reno, Nevada -.

6. Joh, C.Y. and Y.J. Kim, 2004. [Computational aerodynamic analysis of airfoils for WIG (airfoil-in-ground-effect)-craft]. J. Korean Soc. Aeronaut. Space Sci., 32: 37-46, (In Korean).

7. Kikuchi, K., F. Motoe and M. Yanagizawa, 1997. Numerical simulation of the ground effect using the boundary element method. Int. J. Numer. Methods Fluids, 25: 1043-1056.

8. Kim, H.J. and H.H. Chun, 1998. Design of 2-dimensional WIG section by a nonlinear optimization method. J. Soc. Naval Archit. Korea, 35: 50-59.

9. Kim, H.J., H.H. Chun and K.H. Jung, 2009. Aeronumeric optimal design of a wing-in-ground-effect craft. J. Mar. Sci. Technol., 14: 39-50.

10. Kornev, N. and K. Matveev, 2003. Complex numerical modeling of dynamics and crashes of wing-in-ground vehicles. Proceedings of the 41st AIAA Aerospace Sciences Meeting and Exhibition, January 6-9, 2003, Reno, Nevada

11. Lee, S.H. and J. Lee, 2011. Optimization of three-dimensional wings in ground effect using multiobjective genetic algorithm. J. Aircraft, 48: 1633-1645.

12. Masuda, S. and K. Suzuki, 1991. Simulation of hydrodynamic effects of 2-dimensional WIG moving near the free surface. J. Soc. Naval Archit. Jpn., 170: 83-92.

13. Molina, J. and X. Zhang, 2011. Aerodynamics of a heaving airfoil in ground effect. AIAA J., 49: 1168-1179.

14. Murao, R., S. Seki and N. Tomita, 2005. On a study of a WIG with Propeller-Deflected Slipstream (PDS) by using a radio controlled model. Proceedings of the 8th International Conference on Fast Sea Transportation, June 27-30, 2005, Saint-Petersburg, Russia -.

15. Park, K. and J. Lee, 2008. Influence of endplate on aerodynamic characteristics of low-aspect-ratio wing in ground effect. J. Mech. Sci. Technol., 22: 2578-2589.

16. Rozhdestvensky, K.V., 2006. Wing-in-ground effect vehicles. Prog. Aerospace Sci., 42: 211-283.

17. Tuck, E.O., 1984. A simple one-dimensional theory for air-supported vehicles over water. J. Ship Res., 28: 290-292.

18. Yakhot, V., S.A. Orszag, S. Thangam, T.B. Gatski and C.G. Speziale, 1992. Development of turbulence models for shear flows by a double expansion technique. Phys. Fluids A, 4: 1510-1520.

12

CHAPTER

Performance Assessment of the Waterjet Propulsion System through a Combined Analytical and Numerical Approach

Parviz Ghadimi[1], Roya Shademani[1], Mahdi Yousefi Fard[1]

[1]*Department of Marine Technology, Amirkabir University of Technology, Tehran, Iran*

ABSTRACT

The application of waterjets is rapidly growing and they are increasingly being chosen for propulsion in high-speed crafts. Waterjet as a propulsion system of a vessel is also favorable when it comes to maneuvering, appendage drag, draft and fuel consumption at high speeds. Furthermore, waterjet system has recently gained more credibility for its acceptable efficiency because of the advent of more efficient and large pumps. This type of propulsion system consists of many components working together harmoniously, thus establishing a complex system. A significant problem facing designers when predicting performance of the waterjet is the interaction between the hull and the waterjet. This paper describes the powering performance of a vessel equipped with a waterjet system. The interaction between the hull and waterjet is studied in order to predict the powering characteristics. The work starts with an introduction of the waterjet and a review of its current status in design and analysis. Subsequently, hydrodynamic properties of its components are computed and interactions among them are analyzed. Finally, numerical computation is performed for acquiring pressure distribution by a two dimensional computer code in the suction area of the waterjet inlet to predict the possible occurrence of cavitation for the inlet duct.

Keywords: waterjet, inlet flow, numerical analysis, cavitation, propulsion system

1. INTRODUCTION

In 16'Th century, Toogood and Hays for the first time proposed waterjet propulsion system, as reported by J.S. Carlton [1]. At that period of time, waterjet propulsions were used in high-speed pleasure craft and work boats. However, in recent years, this system has been considered for large high-speed crafts. Accordingly, many huge waterjet units have been used in wide range of ships such as passenger and naval crafts.

The waterjet propulsion is a complex system. On the contrary, the screw propellers are simpler, lighter and more efficient than waterjet system. However, the arrival of more efficient pumps, the necessity for timely delivering the critical commercial cargoes, and the required maneuverability for particular vessels have made the usage of waterjets more attractive.

It is normal to divide this type of propulsion system into a hull and a waterjet. It has been demonstrated that waterjet-hull interaction can affect the overall efficiency more than 20% [2]. Usually, waterjet system is broken down into subsystems and an explicit modular approach is applied to analyze them. In order to assess the interaction between the hull and the jet, a parametric method is used.

In 1980, an early contribution related to the jet-hull interaction is attributed to Etter et al. [3]. A complete review of the existing relations for waterjet-hull performance is presented by Allison [4]. Van Terwisga [2]proposed a parametric propulsion prediction method for the waterjet driven craft. Several numerical methods were devoted to study the flow behavior in every part of waterjet system by many researchers whose works were submitted to the ITT Conference [5].

A complete design method that utilized the numerical scheme for the analysis of system parts has been used by CCDOTT [6]. N.W.H. Bulten [7] used CFD methods to design and analyze the whole jet system. Also, a flush-type waterjet propulsion unit was designed with different inner diameter impellers by Moon et al. [8]. In a related work, analysis of a waterjet axial pump was performed numerically [9]. They investigated the performance of the axial-flow-type waterjet based on the variation of the impeller tip clearance.

The main distinction between the procedure implemented here and the previous methods is that the present method applies the empirical, analytical and numerical methods simultaneously to reach a conceptual design of the waterjet system. By numerical analysis, inlet design parameters that describe model geometry have been specified. Moreover, a simple 2-dimensional inlet model has been used here instead of a complex 3-D geometry. By this method, we have an acceptable propulsion system in a short time. Accordingly, a complete jet system design code has been developed that combines the empirical, analytical and numerical methods. By considering the ship geometry limitations and hydrodynamic properties, this software proposes suitable jet conceptual design parameters. Consequently, by using simple 2-D domain for inlet duct and computed pressure distributions for different inlet angles, an optimum inlet has been proposed which leads to minimum losses and maximum efficiency.

2. BASIC THEORY OF SYSTEM

Waterjet propulsion system has three main components: Inlet, Pump and Nozzle. Figure 1 shows typical waterjet arrangement while Figure 2 demonstrates the idealized profile of the jet system. Sea water enters the system with the velocity V_{in} and leaves it with a different velocity V_{jet}. The mass flow rate of the water through the waterjet is given by

$$\dot{m} = \rho A_{jet} V_{jet}, \tag{1}$$

where A_{jet} and ρ are the area of nozzle outlet and density of water, respectively. The thrust produced by the system is equal to the rate of change of momentum:

$$T = \rho A_{jet} V_{jet} \left(V_{jet} - V_{in} \right). \tag{2}$$

Hence, the effective propulsion power delivered by the system is given by

$$P_E = T V_S = \dot{m} V_S \left(V_{jet} - V_{in} \right). \tag{3}$$

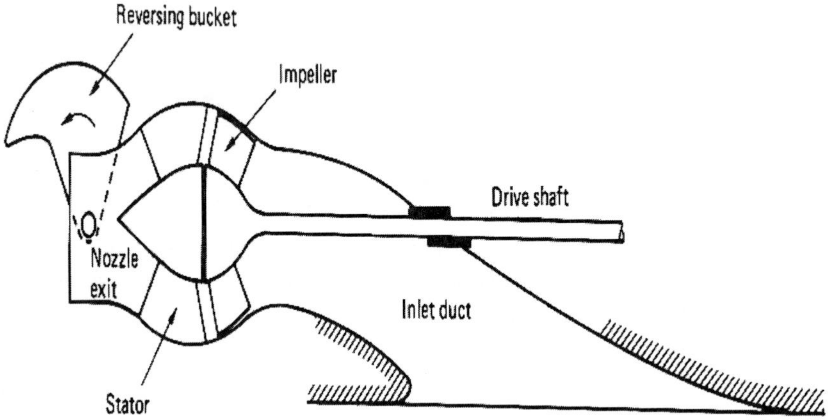

Figure 1. Typical waterjet general arrangement [1]

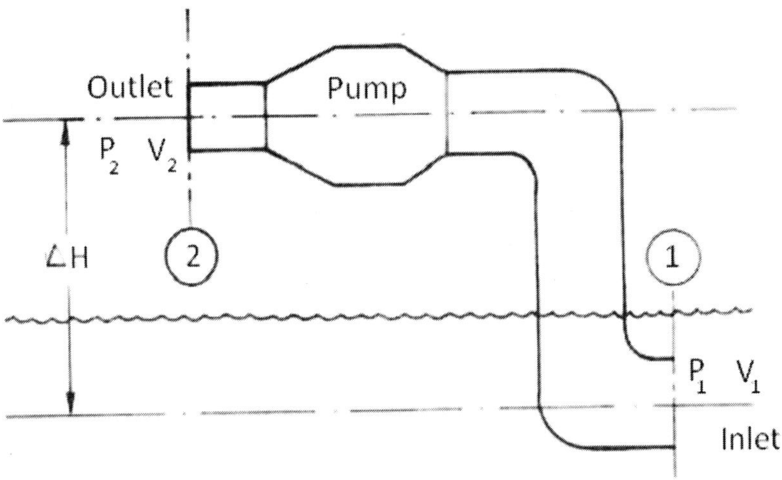

Figure 2. Idealized waterjet arrangement [1]

In Figure 2, the energy equation between points 1 and 2 can be written as

$$\frac{P_1}{\rho g} + \frac{V_1^2}{2g} + H_P = \frac{P_2}{\rho g} + \frac{V_2^2}{2g} + \Delta H + h_{loss}, \tag{4}$$

where

H_P : Pump head

ΔH : Vertical distance between inlet and outlet

h_{loss} : Total head loss.

The jet efficiency is defined as

$$\eta_{jet} = \frac{P_E}{P_E + Losses}. \tag{5}$$

This indicates that the jet efficiency is equal to the ratio of useful power to the total power absorbed by the system.

Because of the effect of hull and geometry of the inlet duct, the inlet velocity is different from ship velocity. Hence, the inlet velocity ratio is defined as follows:

$$IVR = \frac{V_{in}}{V_S}. \tag{6}$$

The inlet has an optimal inlet velocity ratio at which jet has minimum losses at that point. This optimum point has been found by studying a range of velocity ratio and computing related system efficiency. Head losses in the duct inlet and outlet are functions of squared velocity [1] :

$$h_{Inlet} = \xi V_{in}^2 / 2g, \tag{7}$$

$$h_{Outlet} = \psi V_{jet}^2 / 2g, \tag{8}$$

where ζ and ψ are empirical coefficients. These parameters can be assumed as fixed values depending on the inlet and outlet geometries [1] .

Power input to the system (received by the pump) is given by

$$P_{pump} = \dot{m}\left[\frac{1}{2}\left(V_{jet}^2 - V_{in}^2\right) + g\left(h_j + h_{loss}\right)\right], \tag{9}$$

where h_j is the vertical distance between the outlet nozzle and the water line. By separating system losses and applying loss coefficients, we obtain

$$P_{pump} = \frac{\dot{m}}{2}[V_{jet}^2(1+\psi)-(1-\xi)(1-\omega)^2 V_s^2 + 2gh_j]. \tag{10}$$

By using equation (10) and assuming the velocity ratio as $\mu = V_S / V_{jet}$, equation (5) can be written as

$$\eta_j = \frac{2\mu(1-(1-\omega)\mu)}{1+\psi-(1-\xi)(1-\omega)^2\mu^2 + \frac{2gh_j}{V_s^2}\mu^2}. \tag{11}$$

Jet efficiency can be calculated from equation (11). Other design parameters have been found based on the velocity ratios. Figure 3 shows that jet outlet diameter increases at high velocity ratio, while the maximum value of jet efficiency is observed to occur at high velocity ratio. Thus, to achieve high efficiency, a large jet diameter should be selected. However, the aft geometry of ship makes limitations for jet diameter. Therefore, a specified jet diameter is selected and other design parameters are assessed based on this main restriction. On the other hand, the optimum velocity ratio is found by setting the derivative of equation (11) with respect to μ equal to zero to obtain:

$$\mu_{opt} = \frac{1 \pm \sqrt{1-\gamma}}{(1-\omega)\gamma}, \tag{12}$$

where γ is given by

$$\gamma = \frac{(1-\omega)^2(1-\xi) - \frac{2gh_j}{V_S^2}}{(1+\psi)(1-\omega)^2}. \tag{13}$$

By using the pump and hull efficiencies, the total efficiency is defined by the following equation:

$$\eta_{OA} = \eta_{Pump} \cdot \eta_{Hull} \cdot \eta_{jet}, \tag{14}$$

$$\eta_{Hull} = \frac{1-t}{1-\omega}, \tag{15}$$

where η_{Hull}, ω and t are the hull efficiency, wake and thrust deduction factors, respectively. Wake factor can be estimated by the boundary layer theory which will be explained in the next section.

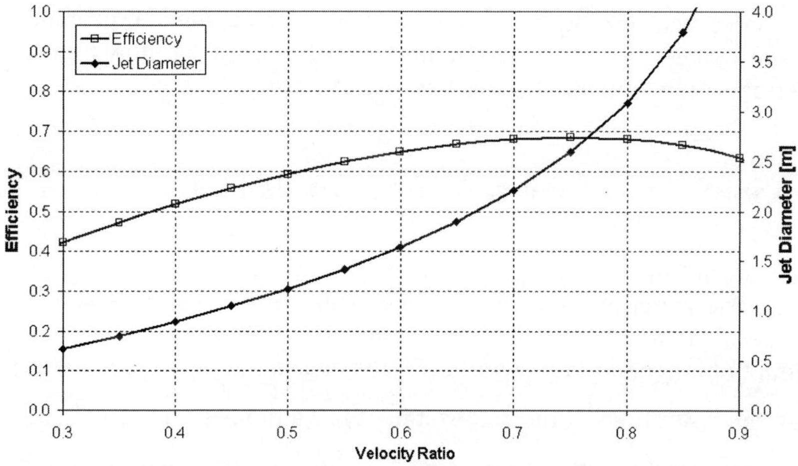

Figure 3. Efficiency and jet diameter as a function of velocity ratio

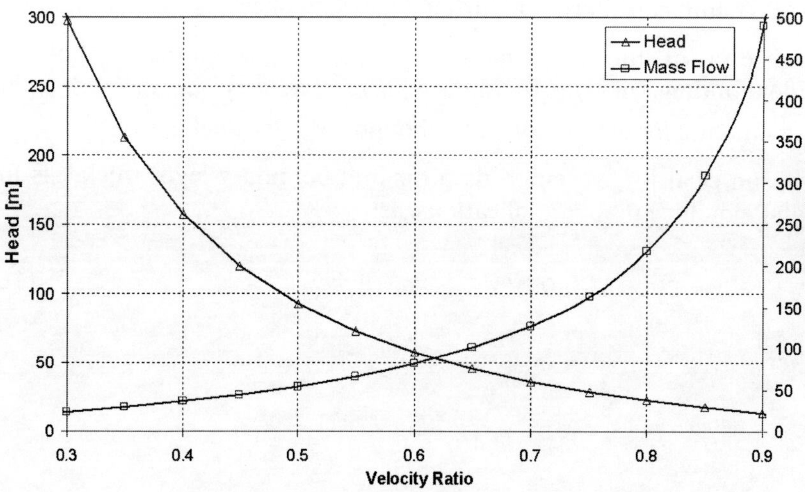

Figure 4. System head and flow rate as a function of velocity ratio 1

Figure 4. shows how the pump head is inversely proportional to the mass flow rate of the jet system. Therefore, for a specified velocity ratio, a pump should be selected as such that its head and mass flow rate correspond to those values obtained from the plots in Figure 4 at that designated velocity ratio. On the other hand, the engine and propulsion system spaces have dimensional limitations, since mixed flow pumps is a better option than other types like axial pumps.

3. ASSESSMENT OF WAKE PARAMETER

The waterjet inlet provides flow from beneath the ship to the waterjet pump. The waterjet inlet is located near the stern area of the ship, and as a result inflow to the flush waterjet inlets will include a significant amount of hull boundary layer flow. This means that for a flush waterjet inlet, the flow entering the waterjet inlet has a momentum velocity, U_m, which is less than the ship speed, V_S, due to the inclusion of hull boundary layer flow. The amount of the hull boundary layer flow which is included in the inlet flow is important as it can impact the size, performance, and propulsive efficiency of the waterjet pump and must be taken into consideration during the design phase of a waterjet pump.

Prediction of the inlet momentum velocity requires a thorough understanding of the boundary layer velocity profile which is instrumental for estimation of the boundary layer thickness, δ.

Svenson et al. [10] presented data for hull boundary layer thickness for high Reynolds number applications, as:

$$\delta = 0.27 \times L_x \times Re_x^{-1/6}, \tag{16}$$

where

$$Re_x = \frac{U_\infty \times L_x}{\nu}. \tag{17}$$

Accordingly, velocity at any point in the boundary layer region is given by:

$$\frac{U}{U_\infty} = \left(\frac{y}{\delta}\right)^{1/n}, \tag{18}$$

where

$$n = \log_{10} Re_x. \tag{19}$$

A standard method of expressing the impact of the inlet momentum velocity is the inlet wake fraction or factor given as:

$$\omega = \left(1 - \frac{U_m}{U_\infty}\right), \tag{20}$$

which can be conversely written as:

$$U_m = (1 - \omega) \times U_\infty, \tag{21}$$

where U_∞, is the uniform speed at far stream (in this case is considered to be the ship-speed V_S) and U_m is the velocity inside the boundary layer (but here, it is the flow speed at the inlet due to the boundary layer). Figure 5 shows the computed values of wake parameter and the boundary layer thickness as functions of ship velocity. It is quite obvious that these two significant parameters decline as ship velocity increases. However, the value of wake parameter is about 0.9 and this parameter could be considered as a constant quantity.

4. VALIDATION OF DESIGN PARAMETERS

As mentioned before, main results of the current code in the form of mass flow rate, pump head and power have been compared with the available results of design software and model test [6]. A range of data made available by the current code was compared with that provided by the CCDOTT software [6]. Furthermore, comparison was made between the results of the current code with the CCDOTT experiment for the design point shown in figures 6, 7, and 8. 'WJ' sign in comparison figures shows the results of the present method. Figure 6 shows the comparison of the mass flow rate for the experimental design point, the current method and fast ferry computed results which was offered by CCDOTT [6]. Also, figures 7 and 8 present the same comparisons for system head and power, respectively. These figures demonstrate that

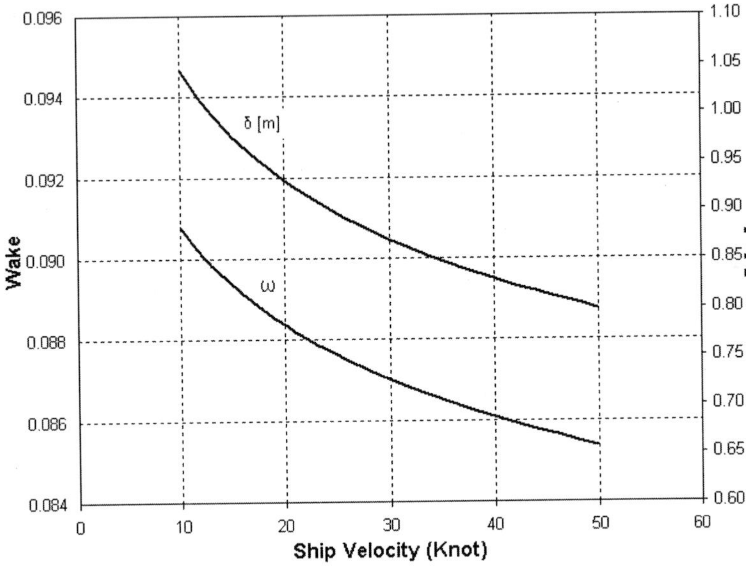

Figure 5. Boundary layer thickness and wake factor as a function of ship velocity

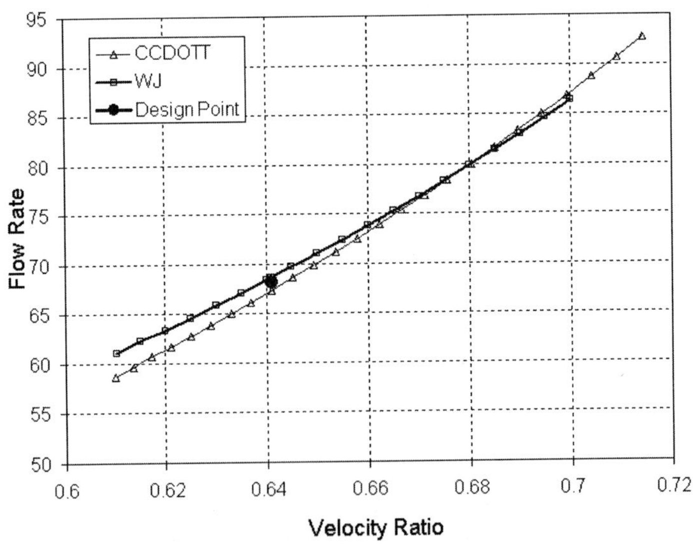

Figure 6. Comparison of flow rate results.

the existing errors between the results of the current code and the experimental data are generally lower than 5%. Based on the presented validations, it can be concluded that the off-design results are quite reliable. The off-design results become particularly important when the high-speed craft tends to reach a maximum speed over the main hump of the speed-resistance curve, because the power which is produced by the engine in off-design condition is sometimes lower than the required value

5. ANALYSIS OF THE INLET DESIGN PARAMETERS

The current code is capable to compute the inlet area and pump diameter from momentum equation and geometry limits. Based on these computed results, empirical theory was used to predict the system losses and basic geometrical parameters. It was found that elliptical inlets have more advantages than the rectangular inlets. Experiments show that elliptical inlet geometry with aspect ratio of 1.3 is the best inlet form for flush type jet [2]. Figure 9 shows the parameterized geometry of the inlet profile. Obviously, a fair inlet profile has many advantages, but sometimes transom geometry and aft form of ship impose limitations. To survey the inlet profile shape and to analyze the flow over this domain, some geometrical parameters have been defined. Cavitation and separation could occur at two points at different inlet velocity ratios [11]. Figures 10 and 11 show the positions of the cavitation and separation phenomena at low and high inlet velocity ratios, respectively. The probability of cavitation occurrence was shown by the current work to be a function of the inlet angle, ramp radius among other design parameters. In the meantime, Inlet angle was found to be the most influential parameter which affects the cavitation occurrence and as such, the best inlet angle was determined. Based on this observation, it was decided that controlling of the cavitation occurrence be done through variation of the inlet angle while keeping other parameters constant.

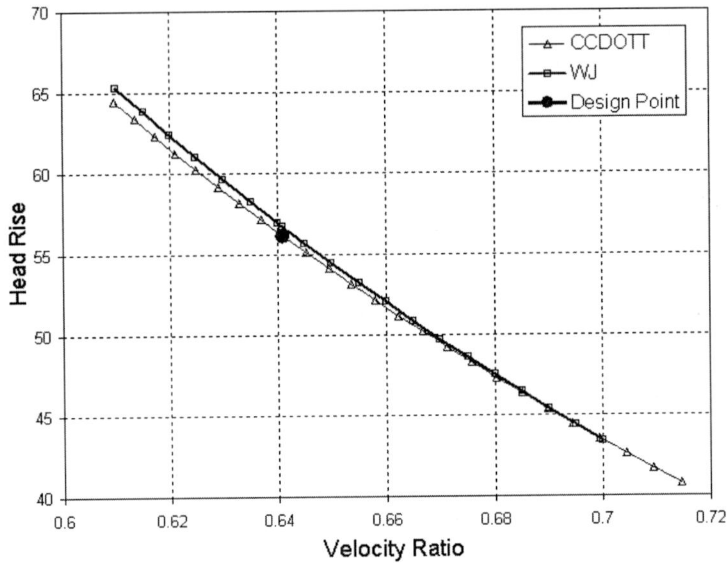

Figure 7. Comparison of system head results

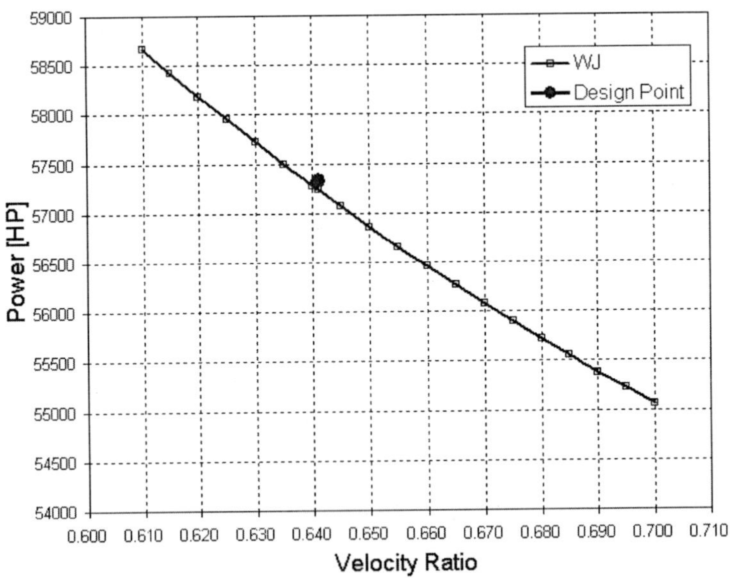

Figure 8. Comparison of system power results

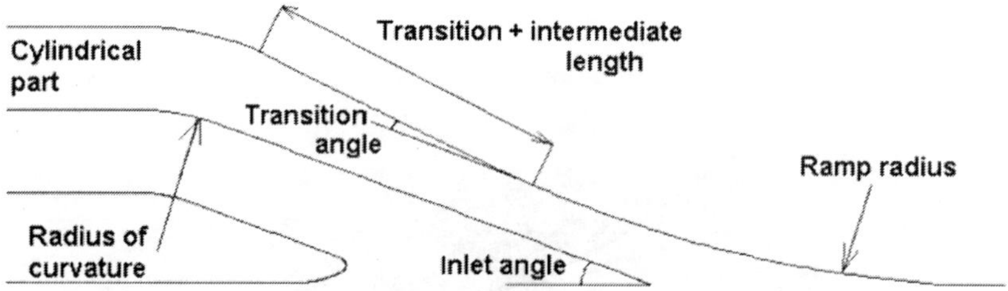

Figure 9. Inlet design parameters

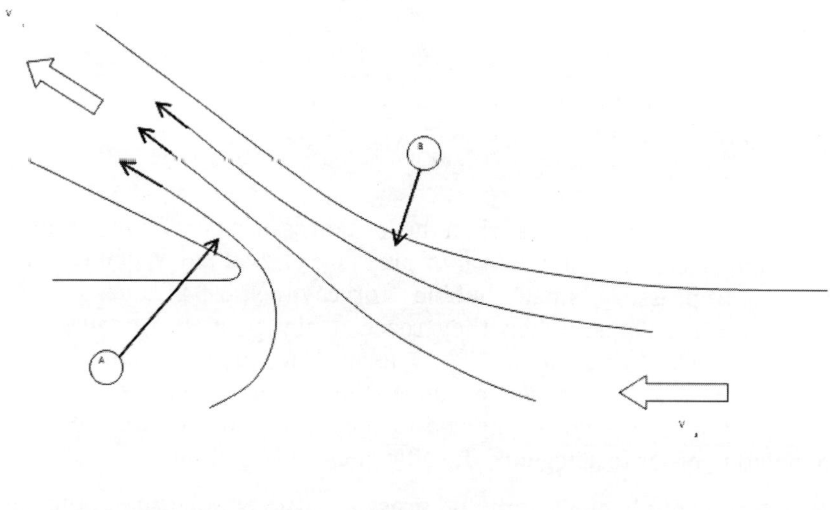

Figure 10. Cavitation (point A) and separation (point B) at low Inlet velocity ratio (IVR)

6. NUMERICAL ANALYSIS OF THE INLET DUCT

When designing a high-speed waterjet inlet, it is difficult to achieve an optimum design that is both efficient and has low drag. This is because cavitation occurs in the water that is unfavorable to both drag and efficiency.

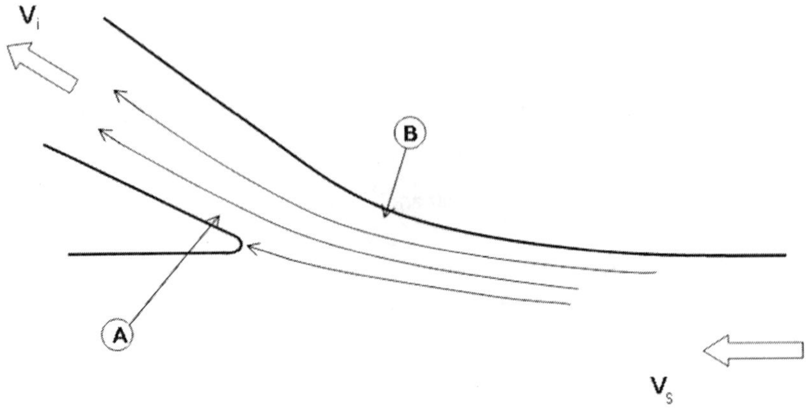

Figure 11. Separation (point A) and cavitation (point B) at high Inlet velocity ratio (IVR)

Cavitation generally occurs when the local pressure on a body moving in a fluid drops to or below the vapor pressure of the fluid. When reaching the vapor pressure, small "bubbles" or cavities are produced. These cavities will collapse when they reach a higher-pressure region and cause a small "water hammer" to form. This phenomenon is called cavitation. Cavitation can produce the negative effects of noise, vibration, and erosion or damage to the inlet, and therefore, must be avoided in order to safeguard the efficiency and the drag.

RANS[2] code has been used to find pressure distribution around the inlet duct. Two-dimensional geometry of waterjet inlet duct has been modeled. Additionally, a rectangular region is constructed around the hull to represent the boundaries of the ocean. Since the ocean is not actually bounded, those boundaries will have to either be placed adequately far away from the ship hull and defined as a wall, or placed closer to the hull to reduce the size of the problem and defined as an opening.

Once the modeling was accomplished, an inlet duct was analyzed for four different inlet angles ranging from 26 to 34 degrees, and minimum pressure at the given inlet angles were compared against the numerical findings of CCDOTT group [6]. This comparison in Figure 12 shows a good agreement between the validation base and numerical calculations. Selection of the inlet angle was done based on two different design considerations; suitable length of propulsion system and avoidance of cavitation. Finding the optimum inlet angle requires a trial and error progress. Accordingly, a large angle was selected and possibility of cavitation was examined, needless to say that assuming a proper safety factor is very important in this progress.

Cavitation number is defined as:

$$\sigma = \frac{p - p_v}{\frac{1}{2}\rho U_\infty^2}. \tag{22}$$

When "$\sigma < 0$", cavitation starts. This implies that high cavitation numbers give less risk for cavitation. Figure 13 shows contours of cavitation numbers in XY plane while Figure 14 presents the cavitation number at the centerline of the system for two different inlet angles. These figures could furthermore indicate how and where the cavitation could occur. Figure 14 shows how the system will operate at a design speed range without cavitation at the inlet ramp when the inlet angle becomes 26 degrees.

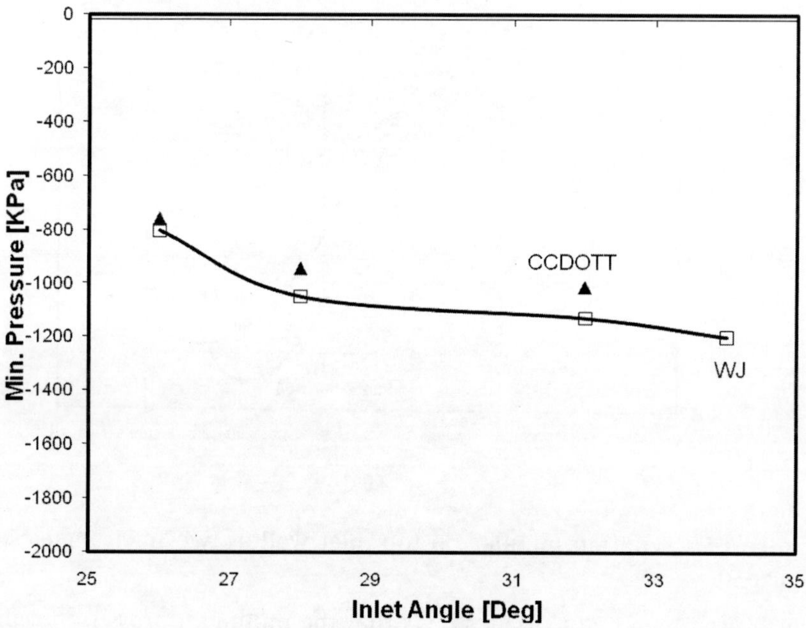

Figure 12. Comparison of minimum pressure for different inlet angles

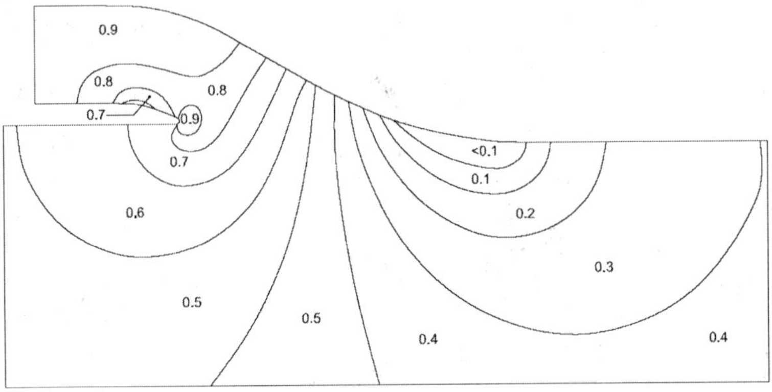

Figure 13. Cavitation number around inlet duct (Inlet angle 26°)

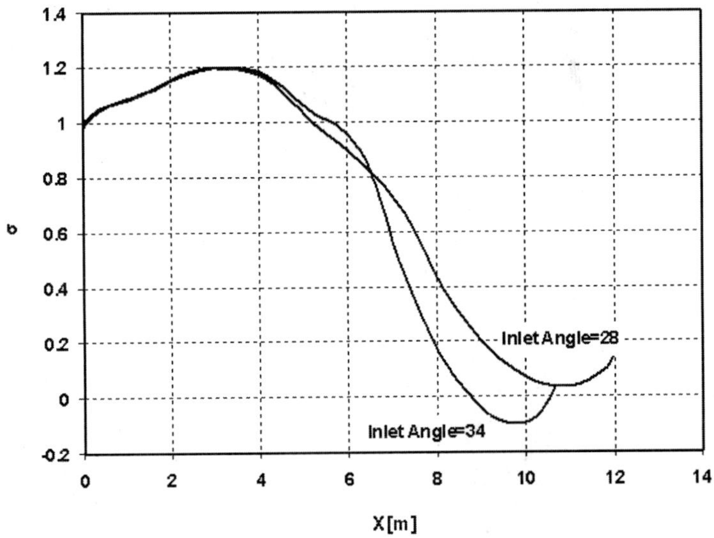

Figure 14. Cavitation number on top inlet wall at two inlet angle (see Figure 9)

which was done by CCDOTT shows that the minimum pressure occurs at the centerline. On the other hand, the results of the current study have good conformity with those of the mentioned 3D analysis. Thus, Figure 14 could be considered as a reliable source for designing the inlet duct.

7. CONCLUSION

Simultaneous accomplishment of the conceptual and basic design as well as the analysis of waterjet propulsion system in a single process is a cumbersome task. Many analytic and empirical methods have been proposed to evaluate the influential parameters of this system. Numerical methods are used for the analysis of the system performance, albeit not in a design process. In the current work, a particular method has been presented which covers all stages of the design and the analysis of waterjet propulsion system which has also been validated by reliable results. Two different codes have been developed, one for the analytical assessment of the main waterjet propulsion system parameters and the other of numerical investigation of the inlet duct impacts.

In the current scheme, a practical method is used to predict the powering characteristics of systems. Prediction of the performance of the waterjet system at the design point starts with determination of the required thrust, jet diameter, shaft horsepower, and RPM which is done by a developed computer program. During the detailed hydrodynamic optimization study, the RPM, radial blade loading distribution, and other parameters of the waterjet were varied to arrive at the optimum design for this power level.

Computed values of flow rate, head rise, and power versus the velocity ratio were compared against earlier reported experimental and numerical results. These comparisons demonstrated good agreements indicating that the separate handling of the waterjet and hull designing process appears to be quite possible. The adopted approach leads to a set of parametric relations that describes the interaction between the hull and the waterjet system. Because of this modular approach, the results can simply be refined during the design process of the vessel.

Empirical relations have been used for preliminary estimation of the required power for the vessel. From this calculated thrust and the assessed internal losses in the waterjet system, the required pump power can be found.

Possible cavitation at the inlet duct leads to erosion or vibrations which must be avoided. In this paper, cavitation characteristics at the inlet were found by looking at the pressure distribution of the water around the inlet opening. Accordingly, the possibility of cavitation was investigated and controlled by close scrutiny of the cavitation number. Along the same line, optimum inlet geometry was found based on the observation of pressure distributions at different inlet angles. This whole process was achieved by a 2-D numerical code written for the flow investigation at the inlet of the waterjet system.

Numerical findings of the current 2-D code compared with the earlier results of the 3-D modeling show an excellent match between the pressure distributions. As a result, one may conclude that a lengthy process of 3-D computation can indeed be avoided and that similar results can easily be achieved by a 2-D analysis which is a less laborious process and saves much computational time.

Main geometrical parameters for the design of waterjet propulsion system have been determined by the present method. By utilizing these important parameters, a designer is able to predict the waterjet performance on a ship before arriving at the basic design stage. Needless to say that there are other considerations that must be taken into account which include necessary number of jets, position of the propulsion system, and the effect of the system on compartmental arrangement of the ship.

NOMENCLATURE

$\dot{m}$: Mass flow rate

ρ: Density

A_{jet} : Area of nozzle outlet

V_{jet} : Jet velocity

T: System thrust

V_{in} : Inlet velocity

P_E : Effective power

V_S : Ship speed

H_P : Pump head

h_{loss} : Total head losses

η_{jet} : Jet efficiency

IVR : Inlet Velocity Ratio

h_{Inlet} : Inlet head losses

h_{Outlet} : Outlet head losses

ξ : Inlet losses coefficient

ψ : Outlet losses coefficient

P_{Pump} : Pump power

h_j : jet height from waterline

ω : Wake factor

μ : Velocity ratio

η_{Pump} : Pump efficiency

η_{Hull} : Hull efficiency

t : Trust reduction factor

δ : Boundary layer thickness

L_X : Distance from bow

Re_x : Reynolds number

υ : Kinematical viscosity

y : Vertical distance from hull

U_∞ : Uniform flow velocity (ship speed)

U_m : Flow speed at the inlet due to boundary layer

σ : Cavitation number

p : Fluid pressure

p_v : Fluid vapor pressure

REFERENCES

1. Carlton.J.S. (John S.), "Marine Propeller and Propulsion", Oxford, Butterworth-Heinemann, 1994T.J.C. Van Terwisga, "A Parametric Propulsion Prediction Method For Waterjet Driven Craft", Fast'97 Conference, paper No. 151, 1997.
2. T.J.C. Van Terwisga, "A Parametric Propulsion Prediction Method For Waterjet Driven Craft", Fast'97 Conference, paper No. 151, 1997.

3. Etter, R.J., Krishnamoorthy, V. and Sherer, J.O., "Model Testing of waterjet Propelled Draft", Proceedings of the 19th ATTC, 1980.

4. Allison, J.L., "Marine Waterjet Propulsion", SNAME Annual meeting, New York, 1993.

5. ITTC, "The Specialist Committee on Waterjets", 22th ITTC, International Towing Tank Conference, 1998.

6. Stanley Wheatley, "Development of a High-speed Sealift Waterjet Propulsion System", Final Report, Center for the Commercial Deployment of Transportation Technologies California State University, Long Beach Foundation, September 30, 2003.

7. Bulten, N.W.H., "Numerical Analysis of Waterjet Propulsion System", PhD Thesis, Technical University of Eindhoven, 2006.

8. Moon-Chan Kim, Ho-Hwan Chun, "Experimental Investigation into the performance of the Axial-Flow-Type Waterjet according to the Variation of Impeller Tip Clearance", Ocean Engineering 34, pp.275-283, 2007.

9. Hong Gao, Wanlai Lim, Zhaohui Du, "Numerical Flow and Performance Analysis of a Waterjet Axial Flow Pump", Ocean Engineering 35, pp.1604-1614, 2008.

10. Svensson, R., etal., "Trial Results Including Wake Measurements from the World's Largest Waterjet Installation", RINA International Conference on Waterjet Propulsion, Amsterdam, 22-23 October 1998.

11. Donald M. MacPherson, "A Universal Parametric Model for Waterjet performance", FAST '99, Aug 1999.

13
CHAPTER

Prediction of Propeller Blade Stress Distribution Through FEA

Kiam Beng Yeo, Wai Heng Choong and Wen Yen Hau

Materials and Minerals Unit, Faculty of Engineering, Universiti Malaysia Sabah, Jalan UMS, Kota Kinabalu, 88400, Sabah, Malaysia

ABSTRACT

The Finite Element Analysis (FEA) of marine propeller blade stress distribution due to hydrodynamic loading is presented and discussed. The analysis provided a better insight to complex marine propeller shape and interaction with hydrodynamic loadings. Stainless steel Wageningen B Series 3 blade propeller with 250 mm diameter, EAR of 0.5 and P/D ratio of 1.2 was adopted in the analysis. The propeller was subjected to the rotational speed of 0-6000 rpm. The pressure distribution demonstrated a positive pressure region on the face section and a negative region on the back section that produces the thrust generation. At 6000 rpm, a maximum positive pressure was achieved at 3225 kPa with a negative pressure of 7229 kPa. The hydrodynamic loading from the pressure distribution computation was applied to the stress distribution computation. From the analysis, the propeller blade stress distribution predicted a highly concentrated region near to the hub and decreasing with the growing value of the propeller radius. The highest stress value of 739 MPa at 6000 rpm was obtained at higher than the stainless steel yield stress (170 MPa) and the blade tip deflected towards the ship hull by 2.73 mm.

INTRODUCTION

In marine propeller design, the propeller blade strength is always an important issue to the cavitation phenomenon for assurance of efficiency and functionality. The propeller blade is developed with an accountable strength based on the material and structural properties serving as the main propulsion component for converting mechanical power to thrust force propelling the ship to move forward through the water. Computational prediction of propeller blade stress distributions due to hydrodynamic loading has been important.

The stress analysis on marine propeller is complex and the solutions accuracy is critical in handling a curvature 3-D model of highly non-linear force, moment and pressure elements. The non-linear force or hydrodynamic loading elements are produced from the thrust and torque and the rotational motion inducing a centrifugal force. Direct analytical modeling of hydrodynamic loading is not a viable solution if realistic prediction is needed. Computational prediction is a significant approach to this problem (Bade and Junglewitz, 2010).

In Vidya Sagar *et al.* (2013) study, FEA based application was adopted for predicting marine propeller under steady state analysis with uniform thrust loading which or is a linear analysis. A clear methodology reported on utilizing the computational method coupling two types of numerical computational (**finite element** method-FEM and boundary element method-BEM) to predict marine propeller performance, hydrodynamic blade loads, cavitation patterns, stress distributions and deflection patterns was conducted by Young (2008). The BEM-FEM had demonstrated the transition of hydrodynamic loading into stress analysis. It provides realistic and accountable results for further application such as optimization of stress concentration region and propeller geometry parameters.

On the complex geometric model adoption in propeller blade, a better solution and shorter computation time for the stress analysis has yet to be reported by researchers. By the thumb of rules, the 2-D model is a much simplified model compared to the 3-D model in computational application. However, the 3-D model had its pros and cons. Chau (2010) stated that the type of numerical solution background theory critically affects the type of geometrical model adoption. Both of the computational 2-D model based on thin shell theory method and the 3-D model using classical theory of deformable solid had shown to produce acceptable similar results.

This study addressed a CFD-FEA prediction of the Wagenigen B Series 3 blade marine propeller pressure and stress distribution subjected to a steady state hydrodynamic loading. The prediction will provide an outcome for further understanding of the pressure and stress distribution on a marine propeller blade.

PROPELLER BLADE STATIC ANALYSIS MODEL

Idealized propeller structure can be simplified as a cantilever beam pivoted at the hub axis with a single loading on the free end or uniformly loaded along the beam. However, this ideal model does not include the highly non-linear wake field or external forces or moment such as the centrifugal forces. As more parameters and flow characteristics with different condition changes, more estimation shall be necessary to improve the effectiveness of theoretical analysis. As the propeller rotates about its central hub axis, each blade suffers different inflow field effect which causes various amplitudes of cyclic resultant moments and forces. Carlton (2007) suggested the general propeller blade stress equation as:

$$\sigma = \sigma_T + \sigma_Q + \sigma_{CBM} + \sigma_{CF} + \sigma_P \quad (1)$$

where, σ_T, σ_Q, σ_{CBM}, σ_{CF} and σ_P are the stress components due to thrust, torque, centrifugal bending, direct centrifugal force and out of plane stress components, respectively.

FINITE ELEMENT ANALYSIS CODE

Finite Element Analysis (FEA) is one of the numerical computational methods in solving or obtaining approximation solution of boundary value problem in linear or non-linear application. The basic fundamental of the FEA is to represent the original model or shape with large number of **finite element** (triangular shape) that permit the finite solution to be carried out on each **finite element**. There are several linear or non-linear analyses that can be carried out by FEA, such as the material **mechanical properties** analysis, dynamic, fatigue, frequency, vibration, thermal and others. Generally, the linear static solution through displacement method in FEA can be described by matrix equation as:

$$[K]\{U\} = \{F\} = \{F^a\} + \{F^r\} \quad (2)$$

where, [K] is the structural stiffness matrix, {U} is the vector of unknown nodal displacement and {F} is load vector ({F^a} and {F^c} of the applied and reaction forces). For {F^a}, it can be redefined to consider the loading as the mechanical {F^m}, thermal {F^{th}} and gravitational load {F^{gr}} and subsequently as:

$$\{F^a\} = \{F^m\} + \{F^{th}\} + \{F^{gr}\} \tag{3}$$

Table 1. Wageningen B series 3 blade propeller with P/D = 1.2

Parameters	Values
Diameter (D)	250 mm
Mean pitch, P_m	300 mm (P/D = 1.
Skew angle (θ_S)	0° (balanced skew
Rake angle (θ)	0°
Propeller Expanded Area Ratio (EAR)	0.5

Then, the mechanical load vector {F^m} is equal to the sum of applied nodal forces and moments and pressure elements as:

$$\{F^m\} = \{F^{nd}\} + \sum_{e=1}^{nel} \{F_e^{pr}\} \tag{4}$$

where, {F^{nd}} is the applied nodal load vector {F_e^{pr}} is the element of pressure load vector, e is the element number and nel is the number of element. Meanwhile the thermal and gravitational load vector can be solved as:

$$\{F^{th}\} = \{F^{nt}\} + \sum_{e=1}^{nel} \{F_e^{th}\} \tag{5}$$

$$\{F^{gr}\} = \left(\sum_{e=1}^{nel} \{M_e\} \right) \{a\} \tag{6}$$

where, {F^{nt}} is the nodal temperature load vector, {F_e^{th}} is the element of thermal load vector, [M_e] is the element of mass matrix and {a} is the acceleration vector.

PROPELLER BLADE MODEL

In this computational analysis, a Wageningen B Series propeller with P/D ratio value of 1.2 had been adopted. The propeller geometrical and particular details are tabulated in Table 1 and the Fig. 1 shows the virtual 3-D propeller model.

FEA BOUNDARY CONDITION AND MESHING

This analysis utilizes the SolidWorks Static Simulation FEA application tool to carry out the propeller blade stress distribution simulation. The propeller blade stresses are mainly due to the medium (water) head pressure and the resultant pressure and force and moment elements due to the geometry iteration with the medium when it been rotated. Computational Fluid Dynamics (CFD) application tool (SolidWorks Flow Simulation) was adopted to obtain the pressure, force and moment elements before transiting to the FEA application tool as external force or moment elements. Each of the propellers simulation underwent the rotating speed of 500, 1000, 2000, 3000, 4000, 5000 and 6000 rpm with non-cavitating flow condition.

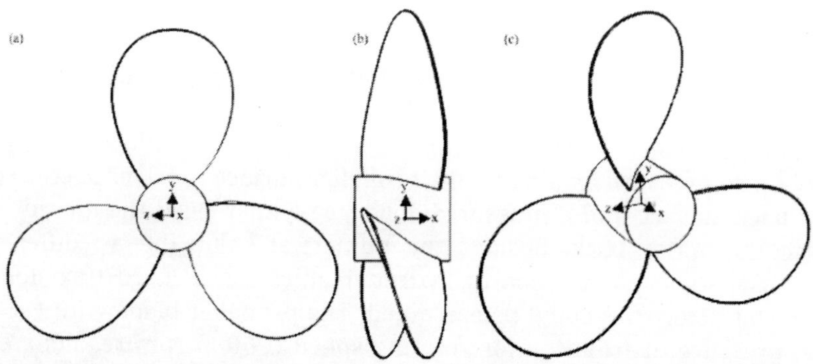

Figure 1(a-c). Wageningen B series 3 blades propeller geometry (a) Front, (b) Side and (c) 3-D view

In order to perform a real world environment, the actual propeller material being assigned as stainless steel with the material properties: Young Modulus, E = 200 GPa, Poison's ratio, NUXY = 0.265, Mass density = 8027 kg m^{-3} and Yield strength = 170 MPa. The propeller model was fixed in the computation domain through the hub surface as shown in the Fig. 2. External loading elements included the pressure, force and moment elements were imported from the previous CFD analysis. The model meshed representing the original model with triangular tetrahedral element model was then generated as shown in Fig. 3. The FEA analysis had been concentrated on the propeller blade stress distribution and displacement due to the hydrodynamic loading.

RESULTS AND DISCUSSION

Based on the Eq. 2-5, the applied load $\{F^a\}$ for propeller blade stress distribution prediction without involving thermal loading through FEA method can be written as:

$$\{F^a\} = \{F^{nd}\} + \sum_{e=1}^{nel} \{F_e^{pr}\} + \left(\sum_{e=1}^{nel} \{M_e\} \right) \{a\} \tag{7}$$

where, the element of pressure load vector $\{F_e{}^{pr}\}$ was preceded from the propeller blade pressure distribution study through CFD application. Wageningen B-Series 3 blade propeller with P/D ratio value of 1.2 was utilized to simulate the blade stress distribution due to the hydrodynamic elements.

Pressure distribution on the propeller blade: From the CFD analysis, each unit of area on propeller solid geometry has its own pressure magnitude which caused the stress distribution variation along the body. The pressure loading on the propeller surface involves suction on the back surface and a pressurized surface which faced the aft side to push the water back. Besides the numerical value due to different loadings, some phenomena in hydrodynamics that caused turbulence flow characteristics could be explained. The propeller blade with P/D = 1.2, pressure distribution for rotating speed 0-6000 rpm results were presented in Fig. 2. Overall pressure distributed on the face section was more uniform compared to the back section as influence by the interaction of hydrofoil geometry profile and the flow stream. The leading edge or hydrofoil stagnation point was subjected to higher pressure compared to the trailing edge. As the propeller rotates, the leading edge cut the flow stream and as a reaction, high pressure was developed at that point and subsequently distributed to the whole blade surrounding. It also shown that the propeller rotation speed greatly

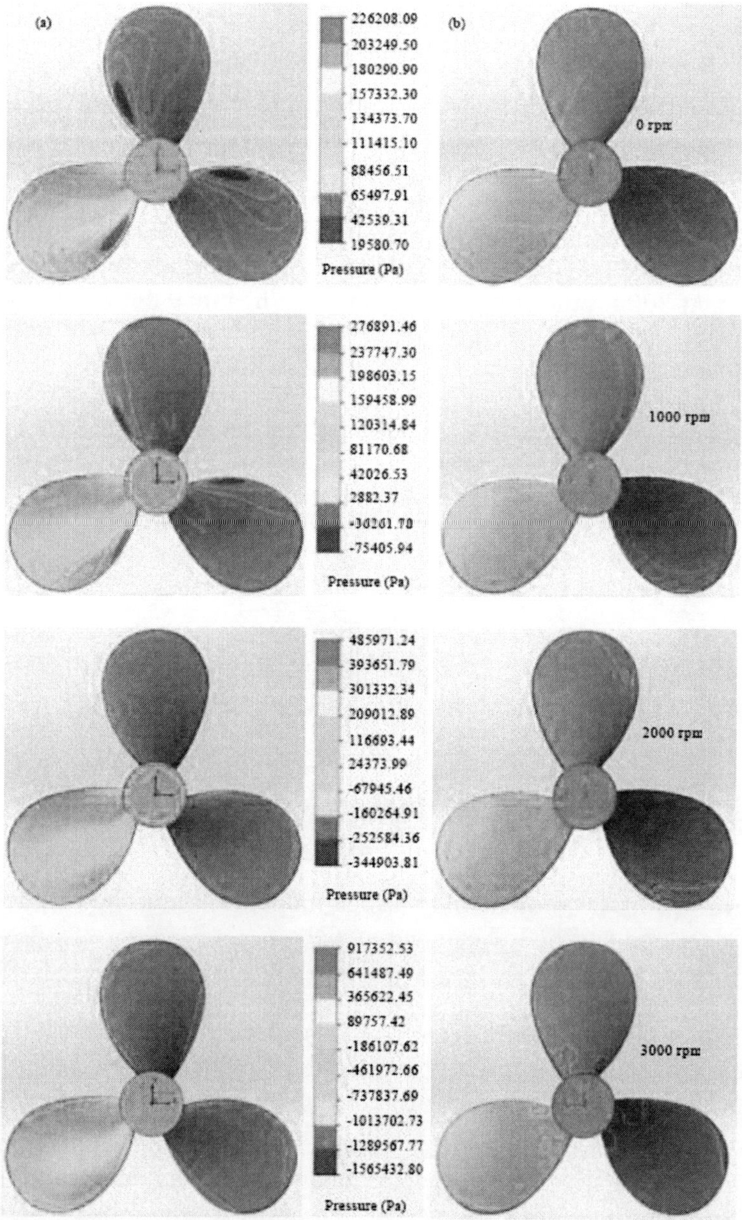

Figure 2(a-b). Propeller blade pressure distribution at various rotational speed (0-6000 rpm). Static analysis on propeller blade due to hydrodynamic loading (a) Back (suction) and (b) Face (pressure)

influences the blade pressure distribution; where the higher the rotation speed, higher the pressure surrounding.

At 0 rpm, the propeller blade was exposed to the water head pressure. The highest pressure region reach 226 kPa at the back section compared to the face region of 103 kPa. The pressure difference highly affected by blade geometry construction. As the propeller reach 1000 rpm, high pressure region was formed at the leading edge. At the back section, a small region of high pressure was formed as a stagnation region. At 2000 rpm rotation, encountering the initial fluid flow velocity boundary condition to produce torque and thrust, high pressure continuously occurred at leading edge with very small region of distribution and very low positive pressure had occurred at the large surface area.

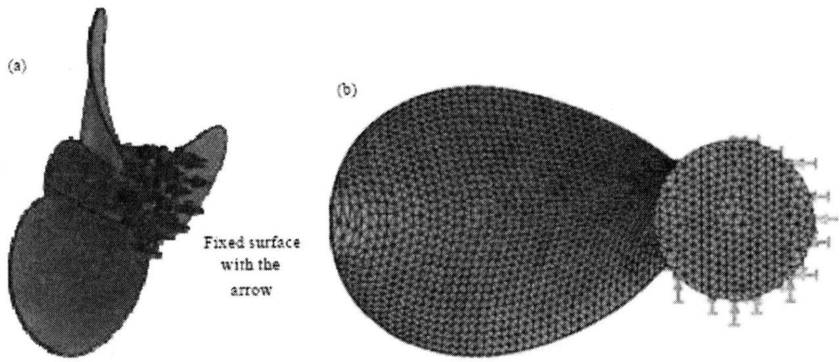

Figure 3(a-b). Unmeshed and meshed model (a) Unmeshed model and (b) Generated tetrahedral element model

A small negative pressure region had also occurred at the back trailing edge. The negative pressure magnitude direction being against the propeller blade satisfied the hydrofoil suction fundamental.

For high rpm performance of marine propeller of 3000-6000 rpm and higher, the negative pressure continued to increase and expand at the back region. At 6000 rpm, positive pressure reached 3225 kPa and negative pressure reached 7229 kPa. The pressure magnitude difference was required for the thrust force generation or forcing the water medium towards to the ship hull for forward movement. This interaction also led to the cavitation phenomenon which required further analysis and validation.

Stress distribution on the propeller blade: Static stress analysis was conducted to predict and verify the propeller blade structure behavior as subjected to hydrodynamic loading or thrust initiated loaded. The analysis is carried out using the SolidWorks Static Simulation FEA application tool.

The obtained propeller blade stress distribution results were shown in Fig. 4.

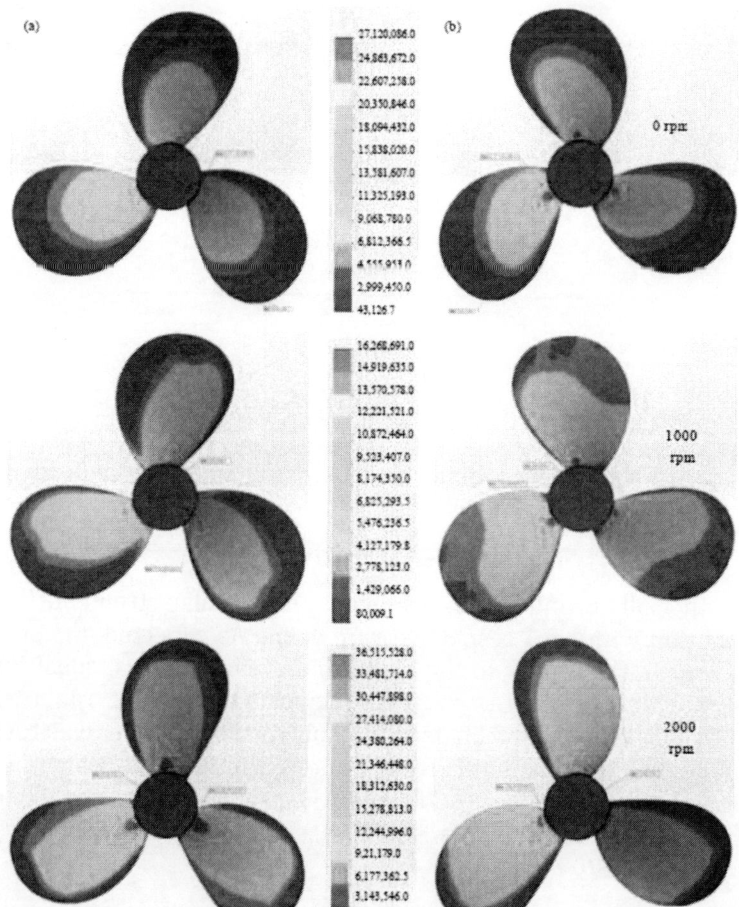

Figure 4(a-b). Propeller blade stress distribution at various rotational speed (0-6000 rpm), (a) Back (suction) and (b) Face (pressure)

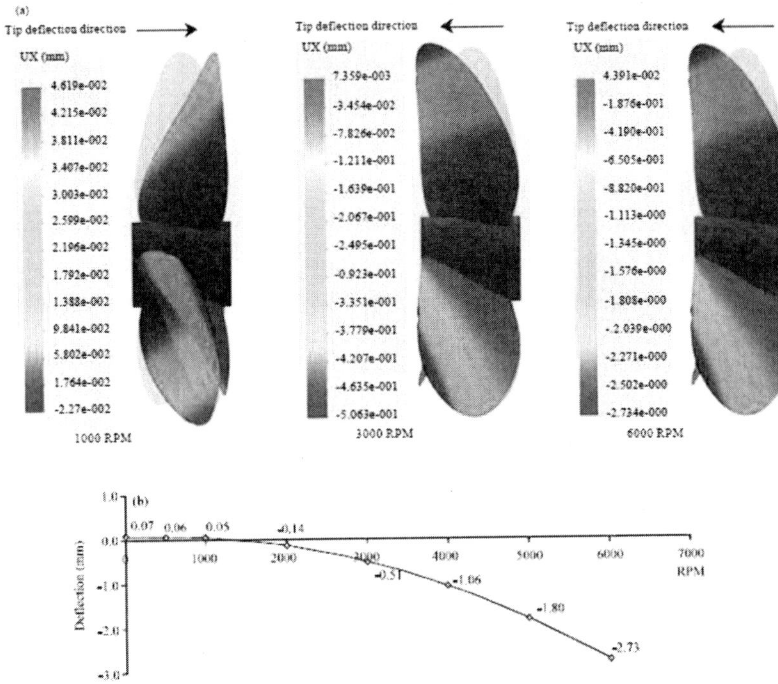

Figure 5(a-b). Propeller blade (a) Displacement due hydrodynamic loading and (b) Deflection direction

Generally, all profile had achieved the similar trend of stress distribution with each respective rpm value. As expected, the stresses were concentrated near to the blade root area and it was ideal for the stresses value to decreases with increasing value of radius as a consequence of fixing the blade at the hub. As the propeller was rotating at 4000 rpm, the maximum stress value reach 285 MPa which further increased up to 739 MPa at 6000 rpm. The propeller will be subjected to face material failure as the stainless steel material yield stress is designed for a 170 MPa.

Deflection distribution on the propeller blade: The propeller blade had deflected due to the hydrodynamic loading as shown in Fig. 5. The maximum deflection occurred at the blade tip free end section. At 0-1000 rpm, there minor tip deformation occurred. As the rpm increases, the tip is deflected towards to the ship hull. The displacement direction concurs with the thrust force and hydrodynamic loading. The degree of deflection had shown to influence by the rpm. At 6000 rpm, tip displacement value reached 2.73 mm. The tip displacement towards the ship hull had increased the pitch value and subsequently allows higher thrust generation. This condition had been a ongoing research focus to develop a composite marine propeller with flexible pitch in-lab.

CONCLUSION

The present analysis demonstrated that marine propeller blade pressure and stress distribution can be predicted through Computational Fluid Dynamic and Finite Element Analysis application tool (SolidWorks Flow and Static Simulation) is significant to marine propeller design and development. The Wagenigen B-Series 3-blade marine propeller with 250 mm diameter and P/D ratio value of 1.2 had been adopted in this analysis. The high pressure and stress concentration regions had been successfully identified for future optimization of propulsion efficiency.

At speed of 2000 rpm, negative pressure form on the back section satisfies the hydrofoil fundamental for thrust generation. The pressure region further developed and at 6000 rpm maximum positive pressure reached 3225 kPa and a negative pressure at 7229 kPa. Cavitation is suitable at low pressure regions of vapor pressure (3.8 kPa at 28°C). Results shown that the propeller is subjected to material failure at 4000 rpm as high stresses (285 MPa and at 6000 rpm is 739 MPa) developed surpassed the stainless steel material yield strength value of 170 MPa. The blade structure displacement had also predicted and the maximum tip displacement value towards the ship hull reached 2.73 mm at 6000 rpm.

ACKNOWLEDGMENT

The authors expressed their appreciation to the Unit Kajian Bahan dan Mineral, Universiti Malaysia Sabah and the Ministry of Higher Education Malaysia for the financial support through the research grant FRG0247-TK-2/2010.

REFERENCES

1. Bade, S.D. and A. Junglewitz, 2010. Automated strength analysis for propeller blades. Proceedings of the 10th International Conference on Computer Applications and Information Technology in the Maritime Industries, May 2-4 2011, Berlin, Germany, pp: 369-378.
2. Carlton, J.S., 2007. Marine Propellers and Propulsion. 2nd Edn., Butterworth-Heinemann, Oxford, UK.
3. Chau, T.B., 2010. 2-D versus 3-D stress analysis of a marine propeller blade. Zeszyty Naukowe Akademii Morskiej w Gdyni, No. 64, July 2010.
4. Vidya Sagar, M., M. Venkaiah and D. Sunil, 2013. Static and dynamic analysis of composite propeller of ship using FEA. Int. J. Eng. Res. Tech., 2: 2587-2594.
5. Young, Y.L., 2008. Fluid-structure interaction analysis of flexible composite marine propellers. J. Fluids Struct., 24: 799-818.

14

CHAPTER

Flexible CFD Meshing Strategy for Prediction of Ship Resistance and Propulsion Performance

Jeong Hwa Seo[1], Dong Myung Seol[1], Ju Hyun Lee[1] and Shin Hyung Rhee[1,2]

Jeong Hwa Seo[1], Dong Myung Seol[1], Ju Hyun Lee[1] and Shin Hyung Rhee[1,2]

[1]Department of Naval Architecture and Ocean Engineering, Seoul National University, Seoul, Korea

[2]Research Institute of Marine Systems Engineering, Seoul National University, Seoul, Korea

ABSTRACT

In the present study, we conducted resistance test, propeller open water test and self-propulsion test for a ship's resistance and propulsion performance, using computational fluid dynamics techniques, where a Reynolds-averaged Navier-Stokes equations solver was employed. For convenience of mesh generation, unstructured meshes were used in the bow and stern region of a ship, where the hull shape is formed of delicate curved surfaces. On the other hand, structured meshes were generated for the middle part of the hull and the rest of the domain, i.e., the region of relatively simple geometry. To facilitate the rotating propeller for propeller open water test and self-propulsion test, a sliding mesh technique was adopted. Free-surface effects were included by employing the volume of fluid method for multi-phase flows. The computational results were validated by comparing with the existing experimental data.

Keywords: Flexible meshing; Resistance performance; Self-propulsion performance; Propeller open water performance; Computational fluid dynamics

INTRODUCTION

Thanks to the great advancement of computer performance recently, computational analysis of ship resistance and propulsion using computational fluid dynamics (CFD) is being widely adopted, and the results are being applied to actual design of ships (Yang et al., 2009; Jasak, 2009; Kim et al., 2010; Seo et al., 2010). In those applications, however, the difficulty in mesh generation is the most serious obstacle for non-expert users for utilizing CFD with maximum efficiency. In case of mesh generation around a ship hull, the bow and stern regions require special care and experience, because of the delicate and rapidly changing surfaces. This is true even for a bare hull and the complexity gets out of control, if propellers, rudders, and appendages are included. Therefore, a hybrid meshing approach using unstructured meshing near the complex geometry and structured meshing in the remaining simple geometry domain was suggested (Lee et al., 2009).

In the present study, we extended the application further and executed resistance and self-propulsion tests using hybrid meshing, which employs unstructured meshing in the bow, stern, and propeller region and structured meshing in the remaining region. For propeller open water and self-propulsion tests, a sliding mesh technique, where the propeller is located in a disk volume and the volume itself rotates with non-conformal interfaces placed between the rotational and stationary sub-domains, was adopted.

The present computational results were investigated through comparison with the experimental datafrom Maritime and Ocean Engineering Research Instituted (MOERI) and other computational results.

COMPUTATIONAL MODEL AND CONDITION

KRISO container ship (KCS), a container ship designed by MOERI, was selected as the object ship, and KP505, the propeller designed for KCS, was selected as the object propeller. The principal particulars are described in Tables 1 and 2.

Table 1. Principal particulars of KCS.

Length between perpendiculars	7.2786m
Beam	1.019m
Draft	0.3418m
Wetted surface area	9.4379 m^2
Speed	2.196m/s
Reynolds number	1.4×10^7

Table 2. Principal particulars of KP505.

Diameter	250.0mm
Ae/A0	0.800
Hub ratio	0.180
No. of blades	5
Blade section	NACA66

COMPUTATIONAL METHOD

Mathematical modeling and Numerical Method

The governing equations are written for the mass and momentum conservation, such that

$$\frac{\partial \rho}{\partial t} + \nabla \cdot (\rho \bar{v}) = 0 \tag{1}$$

$$\frac{\partial \rho}{\partial t}(\rho \overline{v}) + \nabla \cdot (\rho \overline{v}\overline{v}) = -\nabla p + \nabla \cdot \overline{\overline{\tau}} + \rho g \qquad (2)$$

where $\overline{v}$ is the velocity vector in the Cartesian coordinate system, p the static pressure, and $\overline{\overline{\tau}}$ the stress tensor given by

$$\overline{\overline{\tau}} \equiv \mu \left| \left(\nabla \overline{v} + \nabla \overline{v}^{T} \right) - \frac{2}{3} \nabla \cdot vI \right| \qquad (3)$$

Where μ is the molecular viscosity, I the unit tensor, and the second term on the right hand side the effect of volume dilation. Once the Reynolds averaging approach for turbulence modeling is applied, the Navier-Stokes equations can be written in Cartesian tensor form as

$$\frac{\partial}{\partial t}(\rho u_i) + \frac{\partial}{\partial x_j}(\rho u_i u_j)$$
$$= -\frac{\partial p}{\partial x_i} + \frac{\partial}{\partial x_j \partial x_i}\left[\mu \left(\frac{\partial u_i}{\partial x_j} + \frac{\partial u_j}{\partial x_i} - \frac{2}{3}\delta_{ij}\frac{\partial u_i}{\partial x_i} \right) \right] + \frac{\partial}{\partial x_j}\left(-\rho \overline{u'_i} \overline{u'_j} \right) \qquad (4)$$

where δ_{ij} is the Kronecker delta, and $-\rho \overline{u'_i u'_j}$ the Reynolds stresses. These Reynolds stresses must be modeled to close equation (4), i.e., for turbulence closure.

The volume of fluid (VOF) method is employed to handle the free-surface wave flow around the ship. Based on the local volume fraction of the q-th fluid, α_q, the appropriate variables and properties are assigned to each cell within the domain. The tracking of the interfaces between the phases is accomplished by the solution of a continuity equation for the volume fraction of phases. For the q-th phase, this equation has the following form

$$\frac{\partial \alpha_q}{\partial t} + \overline{v} \cdot \nabla \alpha_q(\rho \overline{v}) = 0 \qquad (5)$$

Note that the volume fraction equation is not solved for the primary phase, but based on the following constraint

$$\sum_{q=1}^{n} \alpha_q = 1 \tag{6}$$

A single momentum equation is solved throughout the domain, and the resulting velocity field is shared among the phases. The momentum equation depends on the volume fractions of all phases through the fluid properties, which are determined by the presence of the component phases in each cell, e.g.,

$$\rho = \sum_{q=1}^{n} \alpha_q \rho_q \tag{7}$$

In the case of turbulence quantities, a single set of transport equations is solved, and the turbulence variables are shared by the phases throughout the field.

The computations were carried out using FLUENT v6.3, general purpose CFD software. The governing equations were discretized using implicit unsteady method. The second order accurate upwind scheme was selected for the convection terms, while the second order accurate central scheme was for the diffusion terms. The high resolution interface capturing scheme (HRIC) was used for the discretization of the volume fraction equation near the interface. For the velocity and pressure coupling, the SIMPLEC algorithm was used and the k-turbulence model with a wall function was used for turbulence closure. To confirm the convergence, averaged residuals scaled by the initial imbalance of equations were checked. The criterion for the residual was 10^{-6}. In addition, the constancy of the total force on the ship or propeller was chosen as another standard for confirming the convergence.

Computational Mesh and Boundary Conditionsfor Resistance Test

The computational domain is shown in Fig. 1 with its dimensions. In order to accommodate the free surface motion, there is vertical room of 0.047L between the top boundary and initial water surface.

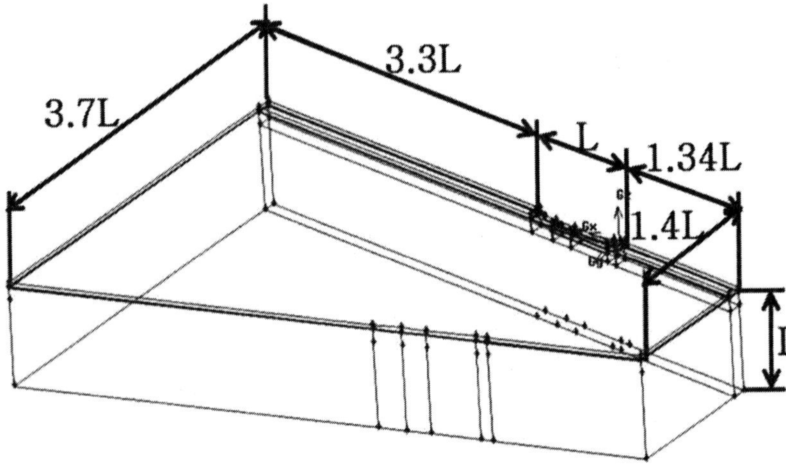

Figure 1. Computational domain for resistance test.

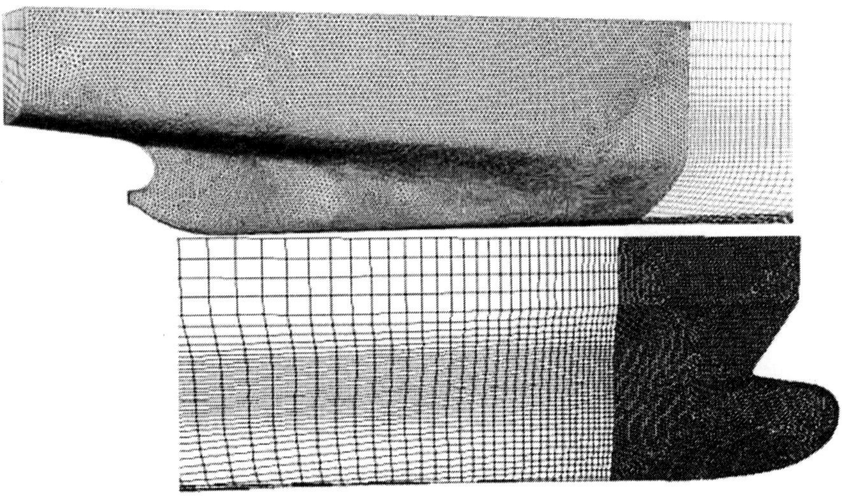

Figure 2. Hull surface mesh.

The no-slip condition was applied on the hull surface. On the upstream and side boundaries, the total pressure including hydrostatic one was applied, while the static pressure was extrapolated through the downstream boundary. The top and bottom boundaries were considered as free slip wall.

Fig. 2 shows the mesh on the hull surface. Unstructured meshing was used in the bow and stern region, while structured meshing was applied in the middle part of the hull. Between the unstructured and structured mesh sub-domains, non-conformal interfaces were placed. The total number of cells used for resistance test was about 1.56 *million,* of which 970,000 cells were tetrahedrons and 590,000 cells were hexahedrons.

Computational Mesh and Boundary Conditions for Propeller Open Water Test

The computational domain around the propeller and its dimensions are shown in Fig. 3. On the propeller and its hub, no-slip condition was applied. On the inlet boundary, velocity components of uniform stream with the given inflow speed were imposed, while the static pressure was set to zero on the exit boundary. Free slip condition was imposed on the outer cylindrical boundary.

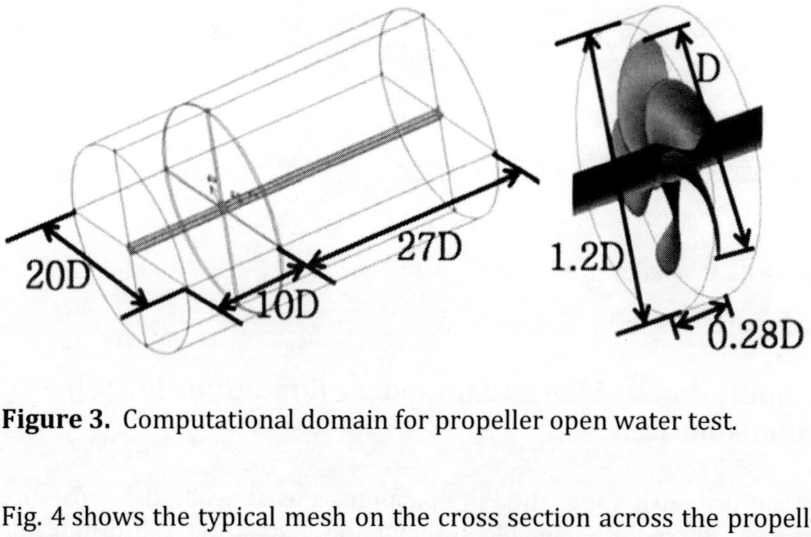

Figure 3. Computational domain for propeller open water test.

Fig. 4 shows the typical mesh on the cross section across the propeller. Immediately around the propeller, unstructured meshing was used, while the outer part is generated by structured meshing. The interface between the structured and unstructured mesh sub-domains is treated as non-conformal interface. To maintain the first cell height off the solid surface between 30 and 120 in terms of y^+, four prism layers are applied on the propeller surface, with 138,012 prismatic cells. The total number of cells used forpropeller open water test was about 1.43 *million,* including 756,000 hexahedron cells and 538,530 tetrahedron cells.

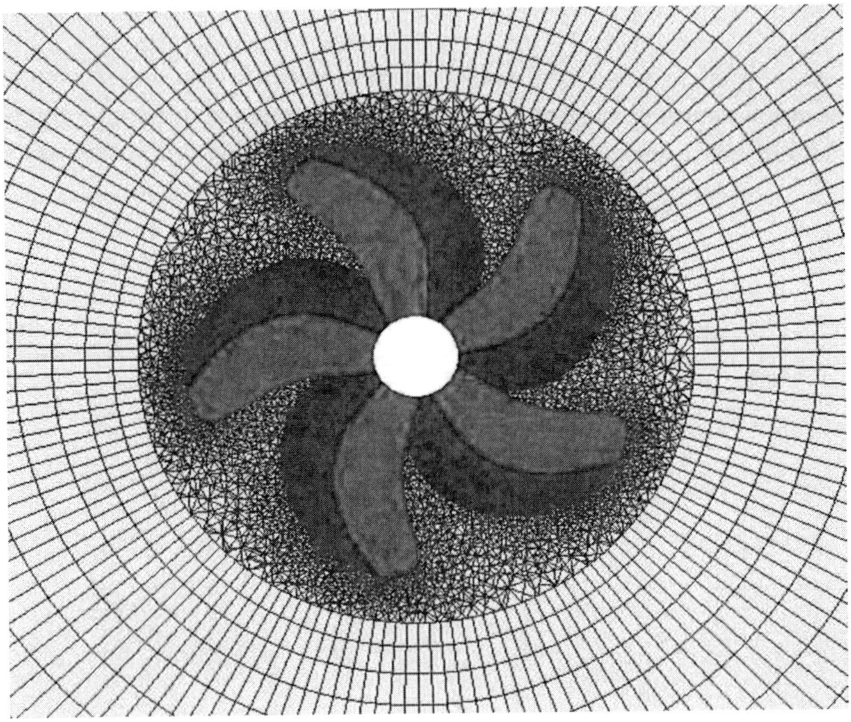

Figure 4. Propeller mesh.

Computational Mesh and Boundary Conditions for Self-Propulsion Test

For self-propulsion test, the full domain was used, while the resistance test used the half domain. Fig. 5 shows the domain for self-propulsion test. Vertical room of 0.047Z for free surface motion exists between the initial water surface and top boundary, as for resistance test. The number of cells for self-propulsion test was about 3.29 *million*tetrahedrons in the bow, stern and propeller regions, and 1.18 *million* hexahedrons in the remaining part. The total number of the cells for the computation was 4.47 *million.*

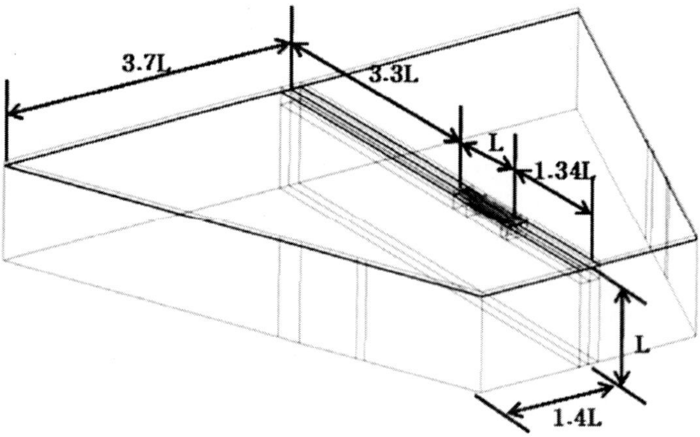

Figure 5. Computational domain for self propulsion test.

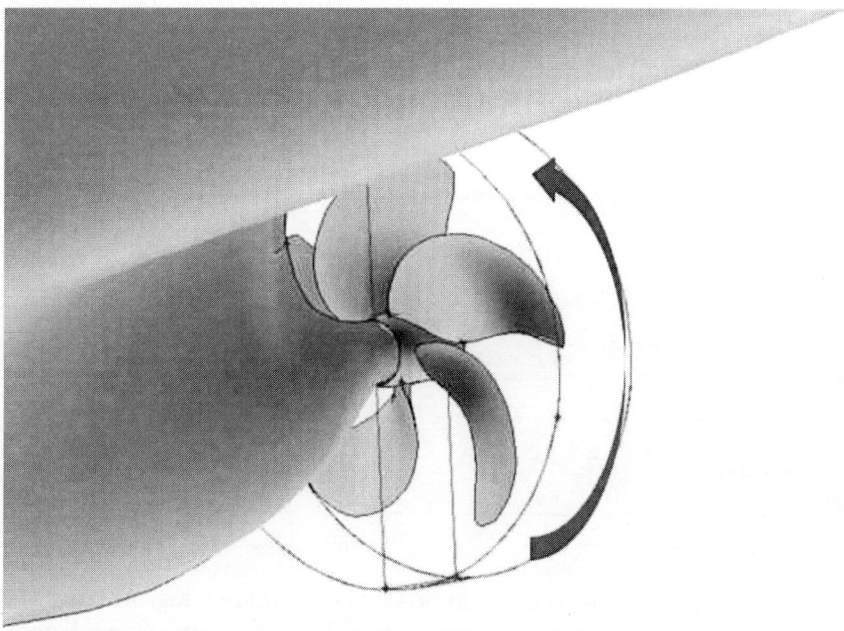

Figure 6. Configuration of sliding mesh.

On the hull surface, no-slip condition applied, while the free slip condition is applied on the top and bottom boundaries. On the upstream and side boundaries, total pressure including hydrostatic one was applied. Through the downstream boundary, the static pressure was extrapolated, as for resistance test. To apply non-conformal interface between the propeller region and hull region, the sliding mesh technique was adopted around the propeller as shown in Fig. 6.

RESULTS AND DISCUSSION

Resistance test

The resistance test results are summarized in Table 3, showing comparison of the resistance coefficients with the experimental data from MOERI and another set of computational results by WAVIS, a CFD code developed at MOERI (Kim et al., 2005). The total resistance coefficient (C_T) and residuary resistance coefficient (C_R) were compared. Note that the frictional resistance coefficient (C_F) was derived from the ITTC 1957 formula. The overall agreement is very good with close comparison with both experimental and other computational results.

Table 3. Comparison of the Resistance Coefficients.

	MOERI (Experiment)	Present	WAVIS
$C_R \times 10^4$	7.316	7.197 (-1.63%)	7.676 (4.92%)
$C_T \times 10^3$	3.560	3.548 (-0.33%)	3.596 (1.01%)

Next, the free-surface wave solutions were compared. Fig. 7 shows the comparison of free-surface wave pattern with the experimental data from MOERI. Although there are irregular contours lines caused by graphical reconstruction of interfaces in tetrahedral cells, the comparison looks quite good, especially near the hull. However, as for other CFD results, the diverging wave away from the hull are under-predicted due to numerical diffusion. Figs. 8 and 9 present the comparison of wave profiles along the hull surface/center-plane and wave cut at y/L_{pp}=0.1509. For this comparison, the computational results from National Maritime Research Institute (NMRI) of Japan were used. It is confirmed that, even across the non-conformal interfaces, the free-surface wave solution shows smooth transition and good agreement with the experimental data. As seen in the wave pattern comparison, numerical diffusion causes under-prediction of free-surface wave solution in both the NMRI's and present results.

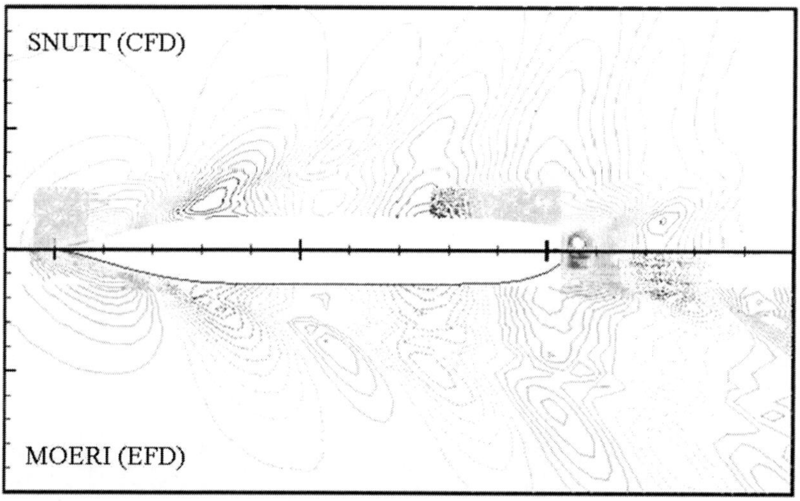

Figure 7. Comparison of Wave Patterns.

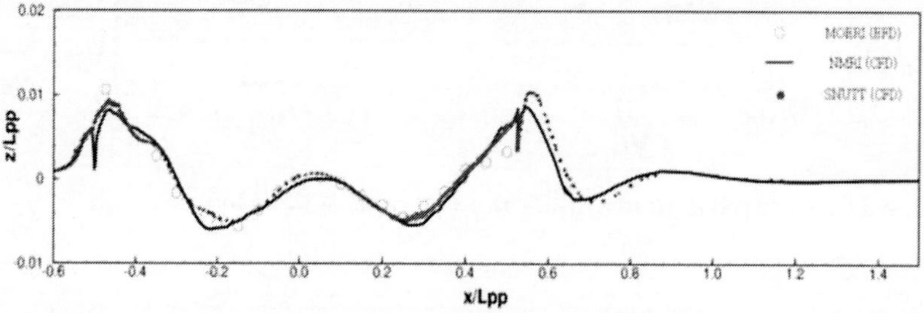

Figure 8. Comparison of Wave profile

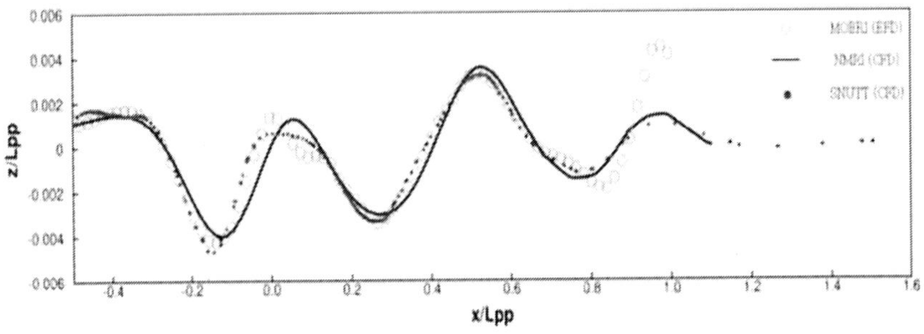

Fig. 9. Comparison of Wave cut at y/L_{pp}=0.1509.

The wake distribution at x/L_{pp}=0.4911(from midship) was compared with the experimental data, as shown in Fig. 10. The overall trend was reproduced well, but because of the tetrahedral cells in the region, the strong longitudinal vortices were slightly under-predicted.

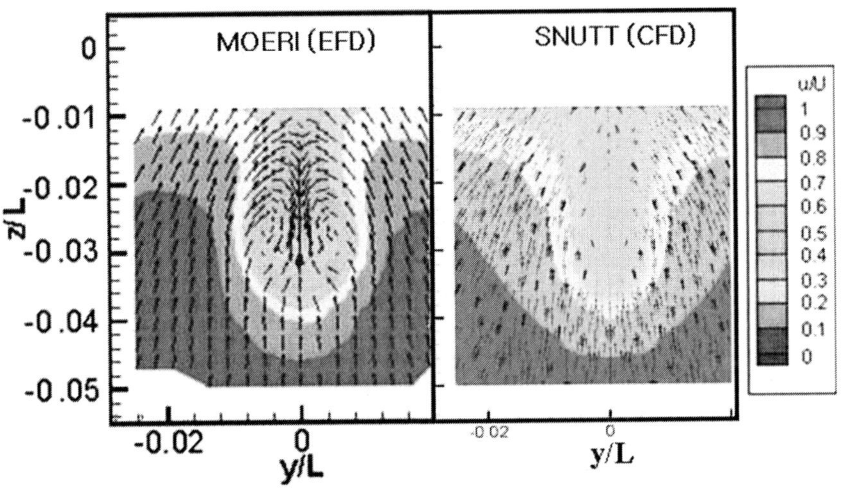

Figure 10. Comparison of wake plane at x/L_{pp} = 0.4911.

Propeller open water test

The computational condition was set to the experimental one, i.e., the propeller revolution at 14*rps* with varying flow speed to set the advance ratio, $J = V/nD$, to 0.1, 0.3, 0.5, 0.7, and 0.9. The range of Reynolds number according to J was from $1.45{\times}10^5$ to $1.30{\times}10^6$. The computed K_T and K_Q values were compared to the experimental data from MOERI, as shown in Tables 4 and 5. The comparison of K_T is better than that of K_Q, and it is believed to be due to the difficulty in accurately predicting the drag on the propeller blade, which is largely responsible for torque. In the course of the computation, it was learned that the wall y^+ value should be less than 100, even with the wall function boundary condition on the blade surfaces.

Table 4. Thrust Coefficient (K_T).

J	Rn	MOERI (Exp.)	Present
0.1	1.45×10^5	0.4763	0.4642(-2.53%)
0.3	4.33×10^5	0.3814	0.3829(0.39%)
0.5	7.22×10^5	0.2763	0.2744 (-0.68%)
0.7	1.01×10^6	0.1770	0.1786 (0.92%)
0.9	1.30×10^6	0.0757	0.0721 (-4.67%)

Table 5. Torque Coefficient (K_Q).

J	Rn	MOERI (Exp.)	Present
0.1	1.45×10^5	0.6720	0.6595 (2.08%)
0.3	4.33×10^5	0.5527	0.5732 (5.05%)
0.5	7.22×10^5	0.4258	0.4583 (3.62%)
0.7	1.01×10^6	0.2992	0.3316 (7.12%)
0.9	1.30×10^6	0.1682	0.1929 (6.30%)

Fig. 11 and Fig. 12 show the surface pressure coefficient distribution at $J = 0.7$ and 0.9, respectively. The general tendency of surface pressure distribution on propeller blade surfaces was well reproduced, and the regions of extremely high and low pressure were clearly identified.

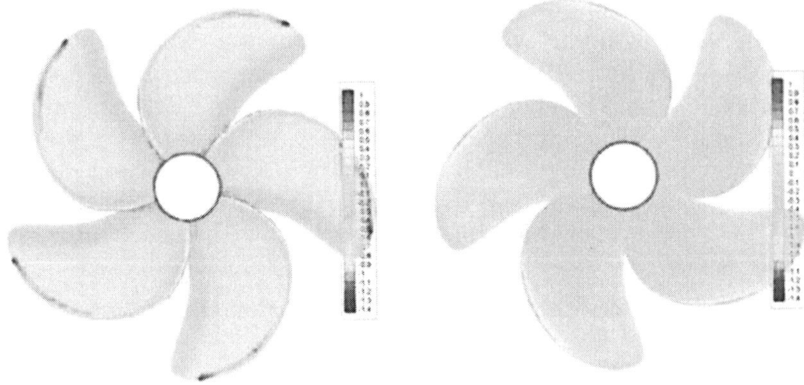

Figure 11. Pressure distributionat J=0.7(left: back side, right: face side).

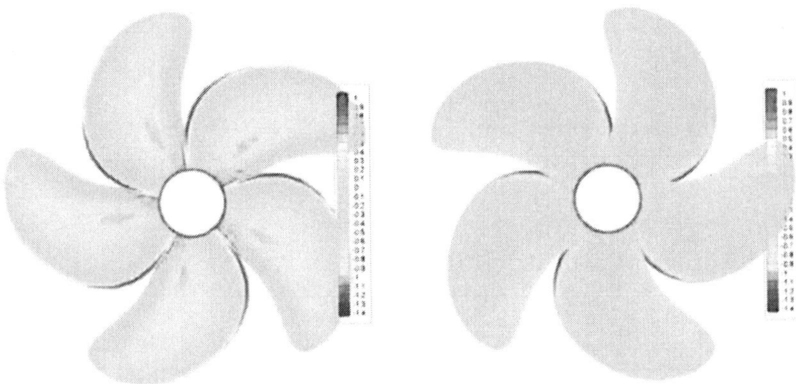

Figure 12. Pressure distribution at J=0.9 (left: back side, right: face side).

Self-propulsion test

It was confirmed, from the above propeller open water test, that the use of the sliding mesh technique can produce reliable solutions for rotating propellers. The same approach was adopted for self-propulsion test with the rotating cylinder placed at the propeller location.

The computation was started with non-rotating propeller to first develop the steady flow field around the hull. Then the propeller was slowly rotated, and the revolution was gradually increased to the desired speed. The time step size was determined to make Courant number no smaller than 0.5. By rough estimate, the propeller revolution speeds of 9.0, 9.5, and 10.0 *rps* were considered, and the self-propulsion point was found by linear curve fitting using those three revolution speeds. For the full scale extension, as for the towing tank tests, the towing force (F_D) was considered to adjust the difference of the viscous force between the model and full scale ships.

The computational results for the three revolution speeds are listed in Table 6, where the error value is defined as $(R_T - T - F_{D0})/0.5\rho U^2 S_M$. Based on these results, the self-propulsion point was searched using a linear interpolation, and the obtained revolution speed, K_T, and K_Q are presented in Table 7.

Table 6. Calculation results of three *rps* cases.

rps	*J*	K_T	K_Q	error
9.0	0.9761	0.1452	0.0281	5.88×10^{-4}
9.5	0.9247	0.1721	0.0307	-9.92×10^{-5}
10.0	0.8785	0.1903	0.0331	-3.95×10^{-4}

Table 7. Result parameters of self propulsion test.

rps	K_T	K_Q
9.592	0.1700	0.03110

The pressure distribution on the hull surface and wake distribution at x/L_{pp}=0.4911 are shown in Fig. 13, Fig. 14 and Fig. 15, with comparison to the experimental data. The pressure recovery around the bilge was well predicted both quantitatively and qualitatively. Also the effects of rotating propeller in the hull wake conform to those observed in the experiment.

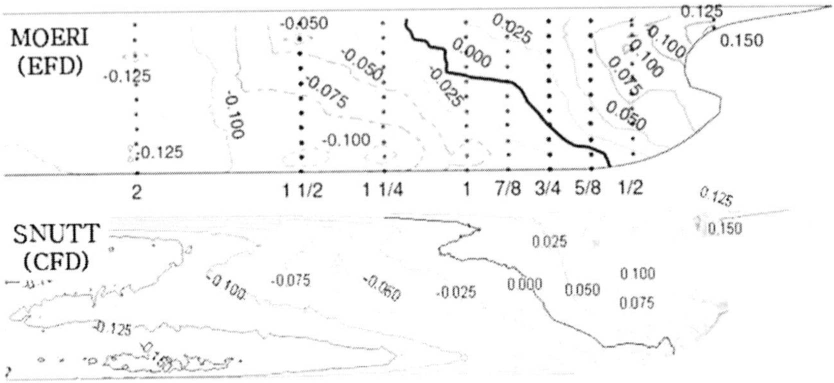

Figure 13. Comparison of hull surface pressure distribution.

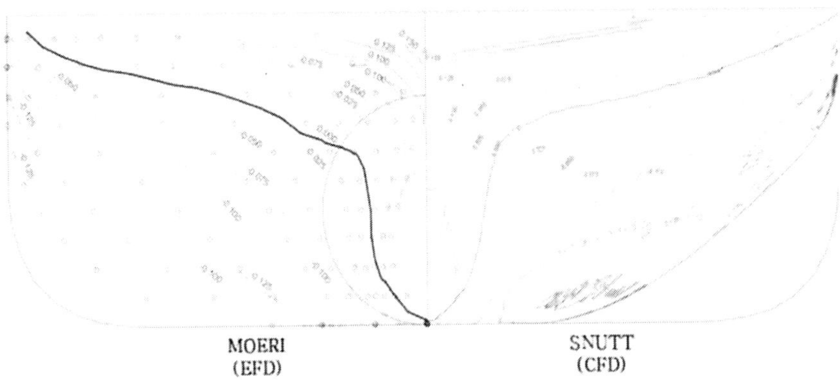

Figure 14. Comparison of pressure distribution(port side).

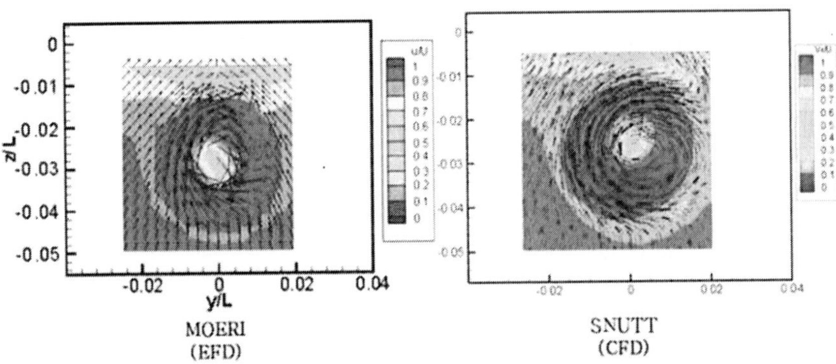

Figure 15. Comparison of wake plane at x/L_{pp}=0.4911.

The pressure distribution on the hull surface and wake distribution at $x/L_{pp}=0.4911$ are shown in Fig. 13, Fig. 14 and Fig. 15, with comparison to the experimental data. The pressure recovery around the bilge was well predicted both quantitatively and qualitatively. Also the effects of rotating propeller in the hull wake conform to those observed in the experiment.

Lastly, the quasi-propulsive coefficient (η_D)was derived using the results from resistance, propeller open water, and self-propulsion tests. Table 8 presents the self-propulsion parameters obtained from the experimental data from NMRI and other computational results reported at the CFD Workshop Tokyo 2005 (Tahara et al., 2005; Karl and Chao, 2005; Lübke, 2005; Kim et al., 2005)

Table 8. Computed self-propulsion parameters.

	$1\text{-}t$	$1\text{-}w_t$	η_0	η_r	J	n	η_D
NMRI (Exp.)	0.835	0.792	0.682	1.011	0.728	9.50	0.740
MOERI	0.846	0.779	0.671	1.023	0.729	9.38	0.746
HSVA	0.865	0.789	0.667	0.981	0.725	9.56	0.717
SVA	0.910	0.765	0.614	1.007	0.708	9.50	0.618
OPU	0.852	0.789	0.631	1.074	0.718	9.53	0.732
Present	0.833	0.773	0.613	1.004	0.708	9.59	0.665

Compared to the experimental data and other computational results, the present results show under-prediction of η_D. It is mainly attributed to the excessive numerical diffusion in the stern region caused by the tetrahedral cells there. However, it should be also noted that most of the other computational results use the body force momentum source terms to represent the propeller effects in self-propulsion test, which must have a significant influence on the thrust deduction factor prediction. It is recommended to refine the mesh generation procedure for the stern region for better prediction of wake and propeller flow there.

CONCLUSIONS

In the present study, flexible meshing techniques, i.e., hybrid meshing for complex geometry regions and sliding mesh for rotating propeller, were employed to perform the three primary tests for ship resistance and propulsion performance, i.e., resistance, propeller open water, self-propulsion tests. Unstructured meshing for the bow and stern regions, together with structured meshing for the remaining region, was used and non-conformal interfaces were placed between the sub-domains with different cell types. The sliding mesh technique, where the propeller was placed in a rotating cylindrical volume with non-conformal interfaces between the rotational and stationary sub-domains, was used to represent the rotating propeller in open water and self-propulsion tests.

The computational results were compared with experimental data and other computational results. The overall performance of the flexible meshing techniques was commendable, but with slight under-predicted wake and quasi-propulsive coefficients. Development of the refinement of mesh generation procedure for the stern region is recommended for future work.

ACKNOWLEDGMENT

This research was supported by National Research Foundation (Grant No. 20090093129) and World Class University project (R32-2009-000-10161-0) funded by the Ministry of Education, Science and Technology of Korea government.

REFERENCES

1. Chao, Y. and Karl, K., 2005. Numeric Propulsion Simulation for the KCS Container Ship. Proceedings of CFD workshop Tokyo 2005, Tokyo, Japan, pp.539-545.
2. Jasak, H., 2009. OpenFOAM: Open Source CFD Research and Industry. International Journal of Naval Architecture and Ocean Engineering, 1(2), pp.89-94.

3. Kim, M.-C. Park, W.-G. Chun, H.-H. and Jung, U.-H., 2010. Comparative Study on the Performance of POD Type Waterjet by Experiment and Computation. International Journal of Naval Architecture and Ocean Engineering, 2(1), pp.1-13.

4. Kim, J. Park, I.R. and Van, S.H., 2005. RANS Computations for KRISO Container Ship and VLCC Tanker using the WAVIS code. Proceedings of CFD workshop Tokyo 2005, Tokyo, Japan, pp.598-603.

5. Lee, J.H. Park, B.J. Seol, D.M. Rhee, S.H. Jun, D.S. Chi, H.R. and Ryu, M.C., 2009. Hybrid Meshing Approach for Resistance Performance Prediction of a POD Propulsion Cruise Ship. Proc. of ASCHT09, 2nd Asian Sympo. on Computational Heat Transfer and Fluid Flow, Jeju, Korea.

6. Lübke, L., 2005. Numerical simulation of the Flow around the propelled KCS. Proceedings of CFD workshop Tokyo 2005, Tokyo, Japan, pp.587-592.

7. Seo, D.W. Lee, S.-H. Kim, H. and Oh, J.K., 2010. A Numerical Study for the Efficacy of Flow Injection on the Diminution of Rudder Cavitation. International Journal of Naval Architecture and Ocean Engineering, 2 (2), pp.104-111.

8. Tahara, Y. Wilson, R. and Carrica, P., 2005. Comparison of Free-Surface Capturing and Tracking Approaches in Application to Modern Container Ship and Prognosis for Extension to Self-Propulsion Simulator. Proceedings of CFD workshop Tokyo 2005, Tokyo, Japan, pp.604- 611.

9. Yang, J. Rhee, S.H. and Kim, H., 2009. Propulsive Performance of a Tanker Hull Form in Damaged Conditions. Ocean Engineering, 36(2), pp.133-144.

Index